Introduction to Ultra Wideband for Wireless Communications

Introduction to Ultra Wideband for Wireless Communications

by

Homayoun Nikookar
TU Delft, The Netherlands

and

Ramjee Prasad
Aalborg University, Denmark
CTiF, Center for TeleInFrastruktur

Springer

Authors

Homayoun Nikookar
Delft University of Technology
Mekelweg 4
2628 CD Delft
The Netherlands
H.Nikookar@tudelft.nl

Ramjee Prasad
Aalborg University
Inst. Electronic Systems
Niels Jernes Vej 12
9220 Aalborg
Denmark
prasad@kom.auc.dk

ISBN: 978-1-4020-6632-0 e-ISBN: 978-1-4020-6633-7

Library of Congress Control Number: 2008932941

Printed on acid-free paper

9 8 7 6 5 4 3 2 1

springer.com

To my loving family Mahsa, Bardia and Parmida.

-Homayoun Nikookar

To my wife Jyoti, to our grand children, Sneha, Ruchika, Akash and Arya

-Ramjee Prasad

Preface

असक्तबुद्धिः सर्वत्र जितात्मा विगतस्पृहः ।
नैष्कर्म्यसिद्धिं परमां संन्यासेनाधिगच्छति ॥

asakta-buddhiḥ sarvatra
jitātmā vigata-spṛhaḥ
naiṣkarmya-siddhiṁ paramāṁ
sannyāsenādhigacchati

Detached by spiritual intelligence from everything controlling the mind, without material desires, one attains the paramount perfection in cessation of reactions by renunciation.

The Bhagvad Gita (18.49)

Compared to traditional carrier-based, Ultra-Wide Band (UWB), or carrier-less, systems implement new paradigms in terms of signal generation and reception. Thus, designing an UWB communication system requires the understanding of how excess bandwidth and very low transmitted powers can be used jointly to provide a reliable radio link. UWB offers systems transceiver potential for very simple implementations.

Comparison between UWB and traditional narrow-band systems highlights the following features:

- Large bandwidth enables very fine time-space resolution for accurate location of the UWB nodes and for distributing network time stamps.
- Very short pulses are effectively counter-fighting the channel effect in very dense multipath environments.
- Data rate (number of pulses transmitted per bit) can be traded with power emission control and distance coverage.
- Very low power density leads to low probability of signal detection and adds security for all the layers of the communication stack.
- Very low power density is obtained through radio regulation emission masks; UWB systems are suitable for coexistence with already deployed narrow-band systems.

UWB systems were born at the end of XIX century stemming from the work of Heinrich Hertz and Guglielmo Marconi. UWB Radio Communication was based on the electromagnetic waves propagating from electrical sparks. The first communication systems were very rudimental and, in a time in which spectral efficiency was not a major concern, the spectrum usage was not efficient.

Starting from spark gap transmissions at the end of the XIX century, research on UWB has been carried on mainly after the Second World War, when sub-nanosecond instruments started to be available. From 1960 until nowadays, continuous research activities in the fields of analog and digital electronics and antennas made it possible to deliver commercial viable UWB systems.

In 2002, in the United States the Federal Communication Commission (FCC) emanated radio-regulation to allow UWB system operations. European

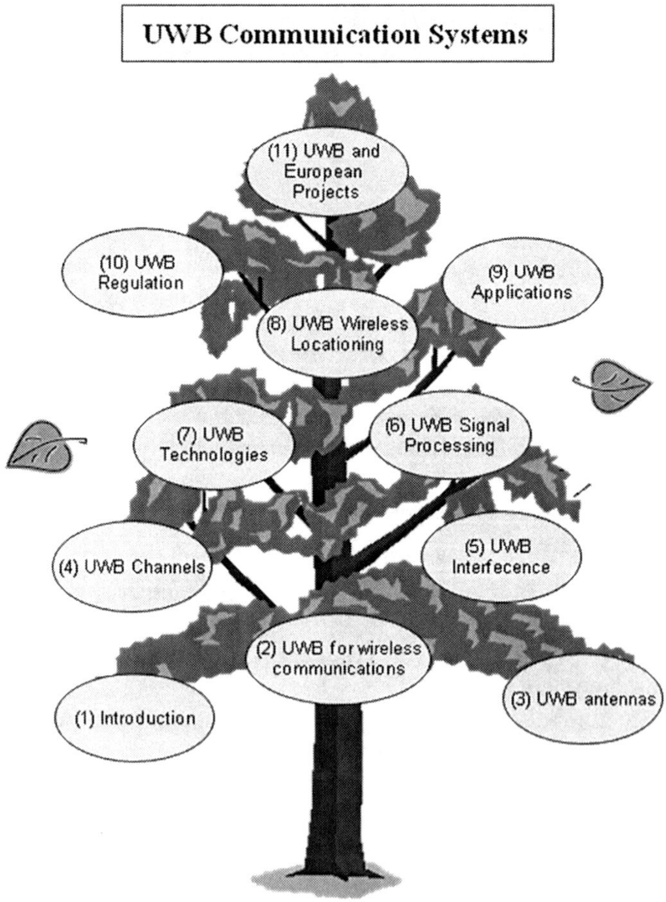

Fig. 1 Tree structure of the book

Commission for Post and Telecommunication (CEPT) has released a decision on the 24th March 2006 which allows operation of unlicensed UWB devices.

UWB technology will support a lot of new applications with benefits for public safety and business creating new opportunities for distributors, vendors and manufacturers. UWB technology will allow an increased usage of spectrum resources by sharing spectrum with low interferences to narrow band systems. The integration of UWB with other wireless access technologies is one of the key factors in order to provide seamless connectivity to next generation mobile user.

The aim of this book is to provide an introduction to the application of UWB to the wireless communications. The book is designed for (graduate) students and researchers working in the field of commercial UWB wireless communications. It will also be useful for the practicing engineers from industry who deal with the wireless systems that are designed and analyzed with the UWB technique. Because of the tutorial nature of the book, it can serve as a textbook on UWB for wireless communications for the Telecommunication graduate program of the electrical engineering curriculum. Problems at the end of each chapter are generated to extend the reader's understanding of the material.

The tree in Fig. 1 illustrates the structure of the book. The book will guide the reader starting from a historical background through all the major design issues of an UWB system. During these years, research activities have been very intense and debated. We would greatly appreciate our readers' comments on this work; collaboration and cooperation will leverage the knowledge of the research community.

Homayoun Nikookar
Ramjee Prasad

Acknowledgements

The material of this book is the outgrowth of research in the Radio Advanced Technologies and Systems (RATS) program of International Research Centre for Telecommunications and Radar (IRCTR) of Delft University of Technology (TUDelft), The Netherlands and Centre for TeleInFrastruktur (CTiF) of Aalborg University, Denmark. Special thanks go to Professor L. Ligthart (Director of IRCTR) for his encouragement and full endorsement of the RATS program and in supporting this book. The authors would like to acknowledge the assistance of M.K. Lakshmanan (a PhD candidate at IRCTR) in making several figures and diagrams and in giving support to the process of writing the book. We also acknowledge the input of Z. Irahhauten (a PhD candidate at IRCTR) and productive discussions with him which has enriched the quality of the book. The assistance of J.R. Nascimento (former exchange student of IRCTR and now with TNO-ICT Delft) is also highly appreciated. The authors acknowledge the fruitful discussions with the members of Dutch Freeband AIRLINK Project and particularly with Prof. I. Niemegeers (from Wireless Mobile Communications WMC Group, TUDelft), Prof. A.Yarovoy (from IRCTR) and Dr. G. Janssen (from WMC Group TUDelft).

Authors would like to further mention valuable contributions of Filippo Meucci and Antonietta Stango (PhDs from CTiF, Aalborg University).

Finally thanks to Junko Prasad for her patience and cooperation in freeing us from the enormous editorial burden.

Contents

About the Authors

Homayoun Nikookar received his BSc and MSc (with distinction) in Electrical Engineering from Sharif University of Technology, Teheran, Iran in 1986 and 1987, respectively and Ph.D. in Electrical Engineering from Delft University of Technology (TUDelft), The Netherlands, in 1995. From 1995 to 1998 he was a post doctoral research fellow at the International Research Centre for Telecommunications and Radar (IRCTR) of the Department of Electrical Engineering, Mathematics and Computer Science of TUDelft, and since 1999 serves as a faculty member of the IRCTR of TUDelft where he is currently an Associate Professor. Dr. Nikookar is also the leader of the Radio Advanced Technologies and Systems (RATS) program of IRCTR, guiding a team of researchers carrying out cutting edge research in the field of radio transmission. He has conducted active research in many areas of wireless communications, including wireless channel modeling, UWB, MIMO, multicarrier transmission, Wavelet-based OFDM and Cognitive Radio. He is the co-recipient of the 2000 IEEE Neal Shepherd Best Propagation Paper Award for publication in the March issue of Transactions on Vehicular Technology. His paper, together with Z. Irahhauten and G. Janssen, (An Overview of Ultra WideBand Indoor Channel Measurements and Modeling, IEEE Microwave and Wireless Components Letters, vol. 14, no. 8, pp. 386-388, Aug. 2004) was listed in the IEEE Xplore Top 100 documents accessed in Feb. and June 2005. He is also recipient of several paper awards of International Conferences and Symposiums. In 2007 Dr. Nikookar served as the Chair of the 14th IEEE Symposium on Communications and Vehicular Technology (SCVT) in Benelux and in 2008 is the Chairman of the European Wireless Technology Conference (EuWiT). He has been technical program committee member of several international conferences including ECWT2003-2008, ICUWB2006, IASTED2007-2008 (Wireless & Optical Communications). He was technical program chairman of European Conference on Wireless Technology in 2004. Dr Nikookar served as a Guest Editor of IEEE Microwave and Wireless Components Letters 2005. He is a senior member of the IEEE.

Ramjee Prasad is Director of Center for Teleinfrastruktur (CTiF), and holds the chair of wireless information and multimedia communications. He is coordinator of European Commission Sixth Framework Integrated Project MAGNET (My personal Adaptive Global NET) Beyond. He was involved in the European ACTS project FRAMES (Future Radio Wideband Multiple Access Systems) as a project leader of TU Delft. He is a project leader of several international, industrially funded projects. He has published over 500 technical papers, contributed to several books, and has authored, coauthored, and edited over twenty books. He has supervised over 50 PhDs and 15 PhDs are working at the moment with him.

Prof. Prasad has served as a member of the advisory and program committees of several IEEE international conferences. He has also presented keynote speeches, and delivered papers and tutorials on WPMC at various universities, technical institutions, and IEEE conferences. He was also a member of the European cooperation in the scientific and technical research (COST-231) project dealing with the evolution of land mobile radio (including personal) communications as an expert for The Netherlands, and he was a member of the COST-259 project. He was the founder and chairman of the IEEE Vehicular Technology/Communications Society Joint Chapter, Benelux Section, and is now the honorary chairman. In addition, Prof. Prasad is the founder of the IEEE Symposium on Communications and Vehicular Technology (SCVT) in the Benelux, and he was the symposium chairman of SCVT'93. Presently, he is the Chairman of IEEE Vehicular Technology, Communications, Information Theory/Aerospace & Electronics Society Joint Chapter, Denmark Section.

In addition, Prof. Prasad is the coordinating editor and Editor-in-Chief of the Springer International Journal on Wireless Personal Communications and a member of the editorial board of other international journals. He was the technical program chairman of the PIMRC'94 International Symposium held in The Hague, The Netherlands, from September 19-23, 1994 and also of the Third Communication Theory Mini-Conference in Conjunction with GLOBECOM'94, held in San Francisco, California, from November 27-30, 1994. He was the conference chairman of the fiftieth IEEE Vehicular Technology Conference and the steering committee chairman of the second International Symposium WPMC, both held in Amsterdam, The Netherlands, from September 19-23, 1999. He was the general chairman of WPMC'01 which was held in Aalborg, Denmark, from September 9-12, 2001, and of the first International Wireless Summit (IWS 2005) held also in Aalborg, Denmark on September 17-22, 2005.

Prof. Prasad was also the founding chairman of the European Center of Excellence in Telecommunications, known as HERMES and now he is the honorary chairman. He is a fellow of IEE, a fellow of IETE, a senior member of IEEE, a member of The Netherlands Electronics and Radio Society (NERG), and a member of IDA (Engineering Society in Denmark). Prof. Prasad is advisor to several multinational companies. He has received several international awards; the latest being the "Telenor Nordic 2005 Research Prize".

Chapter 1
Introduction

Contents

At the end of XIX century, Maxwell's work about electromagnetic waves opened a new era in human history. Heinrich Hertz, a German physicist, was the first to prove Maxwell mathematical theory. In 1894, an Italian scientist, Guglielmo Marconi, obsessed by the idea of a wireless connected-world, started to work in his under-roof laboratory, building up the first radio- communication apparatus.

Besides the first radio links operating in Marconi's estate, located nearby a small city in Italy, Bologna, the first UWB communication system was started in London and was linking two post offices at a distance greater than one mile.

In a short while, spark gap became obsolete and, following the invention of the vacuum tube first, and transistors later, continuous wave transmissions faced on the scene.

The interest in UWB was renewed after the Second World War, when subnanosecond instruments became to be available. Since the last two decades of the millennium, continuous research activity in the fields of analog and digital electronics and antennas made possible commercial viable UWB systems to be delivered. At the end of this chapter, after the historical background of UWB technology, a preview of the book is provided.

1.1 History of UWB

UWB history is generally perceived to start after 1960 with the development of Linear Time Invariant System description via impulse stimula. On the contrary, UWB transmissions history is much longer and goes back to the end of XIX century. At that time, telegraphy was already wide-spread but it was suffering because of the long wired connections which were difficult to be built and maintained, especially in case a river crossing was needed. Transatlantic

H. Nikookar, R. Prasad, *Introduction to Ultra Wideband for Wireless Communications*, Signals and Communication Technology,
DOI 10.1007/978-1-4020-6633-7_1,

cables were settled down using gutta-percha insulation but the maintenance was expensive and time consuming. The history of wireless communications can be considered to start at the end of XIX century with the work carried by Guglielmo Marconi. First wireless transmitters were exploiting spark gaps, resulting in very large bandwidth radio-frequency signals.

From the end of XIX century until nowadays, three eras can be devised in the history of development of UWB systems development:

- pioneering era
- subnanosecond era
- contemporary standardization and commercialization era.

Today, deep technical research is by no means extinguished, especially in the field of efficient receivers and position estimation techniques. We report major dates discussed in the following as well as in Table 1.1.

Table 1.1 UWB Eras

Pioneering Era	
1886	Hertz proofs Maxwell equations, first spark gap transmission
1894	Guglielmo Marconi starts his first laboratory in Italy
1893–1896	Righi develops spark oscillators later used by Marconi
1896	Marconi meets Sir William Preece, first UWB one-mile rooftop link in London
1898	Sir Oliver Lodge, biconical antenna
1901	12th December, Guglielmo Marconi reports first Transatlantic wireless transmission
1902	Valdemar Poulsen invented the Poulsen Arc Transmitter
1906	Lee De Forest invents the "Audion", the first vacuum tube. Spark gap transmissions will be quickly replaced by continuos wave radios.
Subnanosecond Impulses Era	
1939	Philip Carter, conical monopole antenna
1941	Nils E. Lindeblad, Coaxial Horn Element antenna
1960	Henning F. Harmuth, Gerald F. Ross, Kenneth W. Robbins, Paul Van Etten started experimentation with Impulse UWB
1962	Hewlett-Packard, sampling oscilloscope commercialized
1962	Georges Robert Pierre Marié, wideband slot antenna
1972	Kenneth W. Robbins, short pulse Coherent processing tunnel diode ultra wideband receiver replaces sampling oscilloscope
1973	Gerald F. Ross, US Patent 3,728,632: "Transmission and reception system for generating and receiving base-band pulse duration pulse signals without distortion for short base-band communication system"
1985	Henning F. Harmuth, large Current Radiator
1989	U.S. Department of Defense (DoD), Ultra Wide Band term was used for the first time
1994	T.E. McEwan, "Micropower Impulse Radar" (MIR), operating at ultralow power
Standardization and Commercialization Era	
1998	Time Domain Corporation, Commercial Time Modulated Impulse UWB system
2000	Mark A. Barnes, UWB slot antenna

Table 1.1 (continued)

2002	February, US FCC, UWB regulation for data communication, safety and radar applications
2006	24th March, CEPT, ECC Decision of 24 March 2006 amended 6 July 2007 on the harmonised conditions for devices using UWB technology in bands below 10.6 GHz
2006	1st December, CEPT ECC Decision of 1 December 2006 on the harmonise conditions for devices using Ultra-Wideband (UWB) technology with Low Duty Cycle (LDC) in the frequency band 3.4–4.8 GHz
2007	31 August, IEEE 802.15.4a-2007 IEEE Standard for Information Technology – Telecommunications and information exchange between systems – Local and metropolitan area networks – specific requirement Part 15.4: Wireless Medium Access Control (MAC) and Physical Layer (PHY) Specifications for Low-Rate Wireless Personal Area Networks (WPANs)

During the Pioneering Era, late nineteenth century, wireless world was about to start. In 1873 James Clerk Maxwell published his pioneering "Treatise on Electricity and Magnetism" reporting the basis equation for the travelling of the electromagnetic waves. In 1886 the German physicist Heinrich Rudolf Hertz proved Maxwell's concept.

UWB history starts when Hertz solved Maxwell's equation, exactly in 1886. Hertz realized two spark gap generators, each one coupled with an antenna. Producing a spark on the first, a gap was created also on the second generator which was at a certain distance. As a physicist, he was only interested in proving Maxwell's concept and he did not realize the enormous potential of spark gap transmissions.

When Hertz died at the age of 37, his obsession of having a world communicating through wireless links passed to a young Italian boy, Guglielmo Marconi. Born on April 25th 1874 in Bologna, he died in Rome on July 20th 1937. In 1909 he was awarded with the Nobel Prize in Physics thanks to his fundamental contributions to the development of wireless telegraphy.

Marconi, the first wireless communication engineer in the history, got thundered while reading Hertz biography on holiday on Italian Alps in the summer of 1894 at the age of 20.

He was really obsessed by the idea of building a wireless communication system with Hertzian waves. As soon he was back to Villa Grifone estate near Bologna, with the help of his brother, he started a two room laboratory under the roof.

Figure 1.1 shows the first Marconi transmitter. Marconi used this equipment for the first wireless transmission experiment in 1895. The antenna was made by a metal plate and the radiation efficiency was very low. The distance achieved was in the order of hundreds of meters. The radiation was induced by a spark.

Marconi's initial work was based on Righi's spark gap oscillator which is visible in Fig. 1.2. Righi was an Italian physicist who, between 1893 and 1896, developed Hertz's work building a new oscillator able to generate electromagnetic waves which wave-length was just a few centimeters long.

Fig. 1.1 Marconi's first transmitter used for the first wireless experiments in 1895. (Working replica. Marconi Museum, Bigazzi collection. Pontecchio, Italy)

Fig. 1.2 Four spheres Righi oscillator

Marconi's system was based on the following components:

- a spark producing radio transmitter, originally designed by Righi
- a wire or capacity area placed at a height above the ground
- a coherer receiver with a Marconi antenna (a vertical dipole over a ground plane)

The receiver was an upgraded version of Edouard Branly's original device. The signal sent was created by a telegraph key sending short and long pulses, corresponding to the dots-and-dashes of the Morse code. The received signal was recorded on a paper tape.

Years later, in his memories, Marconi reported: "My chief trouble was that the idea was so elementary, so simple in logic, that it seemed difficult for me to believe that no one else had thought of putting it into practice. Surely, I argued, there must be much more mature scientists than myself who had followed the same line of thought and arrived at an almost similar conclusion".

By 1896, Marconi had already built a very efficient apparatus that he could apply for a first patent and seek for funds. Local government shortseeing was not convinced by his invention; Marconi went to England where he got funds by the British Post Office. At the age of 22, Marconi had a meeting with Sir William Preece, the chief telegraph engineer in Britain and one of the most influencing people in the world in the field of communication networks. Preece, which was sixty years old at that time, understood immediately and fully the enormous potential of spark gap transmissions. In two weeks Marconi set up a rooftop link with spark gap transmitters, from General Post Office building in St. Martin's-le-Grand to a second post office building in Queen Victoria Street about one mile away.

Marconi took spark gap transmission from a physicist laboratory to the real world: the first operating UWB radio link was operating in June 1896.

Early history of UWB transmission went ahead with the famous "S" transmission across the Atlantic on 12th December 1901. This first transmission proved the feasibility of the transatlantic wireless link. It was not until 1907 that a reliable wireless link could be set up.

In the first decade of XX century, wireless apparatus started to be quite common and the first problems arose. Spark gap transmission was occupying large part of the radio spectrum, no syntony circuits were available and two stations could not operate at the same time in the same geographical locations. Spark transmitters were heavy and consuming a big amount of energy. Better spark gap generators were under research in those years. An important step ahead was achieved by Valdemar Poulsen (1869–1942), a Danish electrical engineer and inventor of the Poulsen Arc Transmitter in 1902. Until then, high voltage spark gap transmission was the unique way to generate Hertzian waves.

After that, the achievements in the field of vacuum tubes (stemming from "Audion" vacuum tube invention by the American inventor Lee De Forest in 1906) communications were dominated by continuous wave radio transmissions.

A new interest in UWB technology started again with contributions in the late 1960s with works by Henning F. Harmuth at Catholic University of

America, Paul van Etten at the Air Development Center in Rome and Ross and Robbins at Sperry Rand Corporation.

In 1962, Ross begun to describe the response of some microwave networks for the transient regime through their response at an impulse stimulus. At that time, Linear Time-Invariant (LTI) systems where characterised by the more conventional mean of a swept frequency response (i.e., amplitude and phase measurements versus frequency). Ross started to describe an LTI system in terms of its response to an impulsive excitation: h(t). The output signal y(t) to any input x(t) with arbitrary waveform could be uniquely determined via the convolution integral of the input with the impulsive response.

Impulsive technique, however, did not have easy life in that time since the impulse input was not easy to realize due to very quick time requirements. It was not until the advent of the sampling oscilloscope (1962) and the development of subnanosecond pulse generation that the impulse response could be measured and observed directly with sufficient accuracy.

In short time, it became evident that short pulse radar and communications systems could be developed within the same framework used for characterizing LTI system with the impulse stimuli.

Harmuth papers and books published during the years between 1969 and 1984 exposed to the public the basics for UWB transmitters and receivers.

During almost the same period (from 1972 to 1987) and in an independent way, Ross and Robbins started to file patents. They pioneered the use of UWB signals for various applications not only related to communication apparatus, but also concerning radar and sensing applications.

However, UWB communication systems were still lacking sensible receivers. A key turning point was the invention of a short pulse receiver (Robbins 1972) to replace the time-domain oscilloscopes that were large and heavy.

Ross filed a milestone patent to the US Patent office on 17th April 1973, number US 3,728,632: "Transmission and reception system for generating and receiving base-band pulse duration pulse signals without distortion for short base-band communication system". The first modern UWB communication system was born.

Impulse measurement techniques were applied to the design of wideband radiating antenna elements (Ross 1968) and at the Sperry Research Center Ross applied these techniques to various applications in radar and communications (Bennett & Ross 1978).

During the 1980s, UWB technology was referred alternately to as impulse, carrier-free or baseband. The term "UltraWideBand" was first coined by the U.S. Department of Defense in 1989. At that time, UWB theory and many hardware apparatus had experienced almost 30 years of development. Before 1989, Sperry had filed over 50 patents in the field covering UWB receivers, transmitter and pulse generation. Applications covered ranged from radars, communications systems to positioning systems, liquid level sensing, altimetry, vehicle collision avoidance and positioning systems.

In 1994, T.E. McEwan built the "Micro power Impulse Radar" (MIR). This was the first application operating at ultra low power (9V cell operated). It was extremely compact and inexpensive. The radar used quite sophisticated signal detection and reception methods (McEwan, 1994, 2000).

In parallel, fervent research activities have been conducted in the field of UWB antennas. Chapter 3 is specifically dedicated to an analysis of antennas design with ultra-wide bandwidth for short range applications. Antennas are in fact coupling devices between guided waveforms and the free space propagation; new design techniques had to be developed in order to obtain a good performance over a very large bandwidth. In Table 1.1 some important milestones about antenna developments are also reported.

After the great technical developments related to subnanosecond pulses in the years from the sixties until the end of the century [1–7], another rush started with the world-wide activities for technology standardization.

Nowadays, UWB signals can be generated following various paradigms going over the original time-domain Impulse Radio (IR-UWB) transceivers. Cross-pollinated from other research fields in the communication arena, various UWB transmission techniques are illustrated in the following chapters. MultiCarrier UWB (MC-UWB), Orthogonal Frequency Division Multiplexing (OFDM) UWB and Frequency Modulation UWB (FM-UWB) are the strongest candidates for future UWB communication systems.

A major missed milestone in the standardization process was in year 2006. After nearly three years of wrangling over which physical layer should form the basis of an IEEE standard (802.15.3a), the involved parties have given up. At a meeting in Hawaii on January 19, 2006, the IEEE committee 802.15.3a, which was tasked with developing a standard, voted unanimously to disband.

The two opposing groups issued a joint statement on January 20 where they agreed to let the market decide which UWB Physical (PHY) layer will become the "de facto" standard.

One group (UWB Forum) is proposing the direct-sequence UWB while a second, WiMedia Alliance proposes MC-UWB.

However some important standardization activities did not disband and produced their final output and the following were published: IEEE 802.15.4aUWB – Low-Rate Wireless Personal Area Networks (WPANs), Standard ECMA-368 High Rate Ultra Wideband PHY and MAC Standard, Standard ECMA-369 MAC-PHY Interface for ECMA-368, Standard ISO/IEC 26907:2007, Standard ISO/IEC 26908:2007.

1.2 Preview of the Book

In this first chapter, a background from an historical point was discussed. UWB can be considered as the oldest form of radio-communication that was implemented in the history. UWB technology we face today is the results of multi-decade research in sub-nanosecond signal processing, front-end transceivers and antennas design.

In Chapter 2 the use of ultra wideband technique for wireless communications is motivated. Various topics related to the applicability of UWB to wireless systems including the definition of UWB signals, the FCC radiation mask for the UWB transmission, different UWB pulse shapes that can be used as well as the major features of UWB, which are remarkable for wireless communications applications, are studied.

In Chapter 3, UWB antennas for wireless communication are studied. The design and development of antennas for UWB wireless communication is a key research area. The huge bandwidth of UWB systems poses unique research challenges which have to be dexterously addressed. In this chapter a brief introduction to the application of UWB antennas especially for wireless communications is provided.

In order to ensure effective transmission of UWB signals it is important to understand and develop a model of the channel that adequately describes the UWB environment. Developing of a channel model for the UWB environments is very challenging, particularly so because of the very large bandwidth of UWB transmission. The knowledge of the channel helps to a huge extent in the proper design of wireless communication system and evaluation of performance. Chapter 4 focuses on the UWB wireless channel. The important parameters including amplitude fading, time delay, RMS delay spread, based on which the channel models are devised, are explained. The material and results of this chapter are important to the designer of UWB communication system to predict the signal coverage, to estimate the maximum achievable data rate, to determine optimum location for antennas, to design efficient modulation schemes and to study associated signal processing algorithms.

Due the huge bandwidth of the UWB transmission, the degradation in performance of UWB communication because of interference from co-existing services is a major issue. The focus of the Chapter 5 is to study the effects of interference to and from UWB systems. In this chapter, first the general method of signal to interference calculation is explained. Then, in addition to interference of UWB to narrowband systems, the effects of UWB interference on the BER performance of WLAN OFDM system as well as WiMax system are studied and analyzed. Further in this chapter, the effect of narrowband and wideband interferences on the victim UWB system is studied and methods to mitigate the interference are discussed.

In Chapter 6 various modulation schemes of impulse radio UWB are presented. This includes UWB data and multiple access modulations, reception techniques such as Rake and Transmit-Reference receivers. Special emphasis is given to the performances of M-ary pulse amplitude modulation and Pulse position modulation.

In Chapter 7 technologies for UWB transmission are studied. Basically, there are two competing technologies for the UWB wireless communications, namely: Impulse Radio and Multi-band OFDM (MB-OFDM). In this chapter these technologies are discussed in more detail and major characteristics of each technique are contemplated. Further in this chapter, the two technologies are compared from different aspects such as channel, interference, performance and complexity point of view.

UWB technology is especially suited for radio locationing because of its huge bandwidth which provides a fine accuracy in ranging. UWB provides low-power and low-cost communication and positioning in one technology. These features allow a new range of applications, including logistics, security applications, medical applications as well as military applications. In Chapter 8 we study various radio localization techniques giving significant emphasis to UWB wireless locationing. These techniques are based on one or more measurement types such as angle of arrival, time of arrival or time difference of arrival and received signal strength. In this chapter special attention is paid to coherent time of arrival technique (such as ESPRIT, CLEAN, Inverse Filtering, super resolution) as well as the non-coherent method. The problem of multipath propagation and non-line-of-sight location error is explained and the locationing capability of OFDM technique is studied.

The UWB technology can spawn a wide range of interesting wireless applications. In Chapter 9 we list and give a brief account of few of the UWB wireless applications. The key properties of UWB that prompt their applicability to diverse requirements are: high data rate communication, robustness against fading, immunity to multipath, multiple access capability, low cost transceivers and accurate positioning.

In Chapter 10 we see the efforts of various regulatory bodies across the globe to establish standards and norms towards UWB spectrum usage. In this chapter we discuss how the regulations, established as "mask", set out upper limits on the amount of power that can be radiated at any particular frequency, both within and outside the core band of 3.1–10.6 GHz. Further in this chapter it is emphasized that devising a generic regulatory standard that caters to all markets across the globe will be one of the goals of the future for the UWB communications.

At the end of the book, in Chapter 11, we report the major European projects that have been working intensively on UWB system for Personal Area Networks (PANs). Integrated with other wireless technologies, UWB will be a key factor to support a seamless connectivity of the mobile user. Main objective was to define the best Physical Layer (PHY) solution for PAN in terms of performance and interference to existing wireless systems. The reader can go through this chapter and find technical analysis of the PHY solutions presented in the previous chapters. These projects clearly show that UWB is ready to make the entrance in the wireless technologies arena for low-medium and high data rate for medium and short connection ranges.

References

1. C.L. Bennett and G.F. Ross, "Time-Domain Electromagnetics and Its Applications (Invited Paper)", Proceedings of the IEEE, vol. 66, no. 3, pp. 299–318, March 1978.
2. M.Z. Win and R.A. Scholtz, "Impulse radio: how it works", Communications Letters, IEEE. Feb 1998.
3. J.F.M. Gerrits, M.H.L. Kouwenhoven, P.R. van der Meer, J.R. Farserotu and J.R. Long, "Principles and Limitations of Ultra-Wideband FM Communications Systems", EURASIP Journal on Applied Signal Processing, Special Issue on UWB-STATE OF THE ART, vol. 2005, no. 3, pp. 382 – 396, 1 March 2005.
4. I. Oppermann, M. Hamalainen and J.Iinatti, UWB Theory and Applications, Wiley, 2004.
5. K. Siwiak, D. McKeown, "Ultra-Wideband Radio Technology", Wiley, 2004.
6. W.P. Siriwongpairat, K.J.R. Liu, "Ultra-Wideband Communications Systems: Multiband OFDM Approach", Wiley, 2007.
7. M.-G. Di Benedetto, T. Kaiser, A.F. Molisch, I. Oppermann, C. Politano, and D. Porcino, "UWB Communication Systems A Comprehensive Overview", EURASIP Book Series on Signal Processing and Communications, vol. 5, 2006.

Chapter 2
UWB For Wireless Communications

Contents

Ultra wideband (UWB) communication is based on the transmission of very short pulses with relatively low energy. This technology may see increased use in the field of wireless communications and ranging in the near future. UWB technique has a fine time resolution which makes it a technology appropriate for accurate ranging. Because of the huge bandwidth, UWB waves have a good material penetration capability. As will be explained later in more detail in this chapter, the UWB radio signal occupies a bandwidth of more than 500 MHz or a fractional bandwidth of larger than 0.20, [1]. According to Shannon's capacity formula, this large bandwidth provides a very high capacity. Thus, high processing gains can be achieved that allow the access of a large number of users to the system. The impulse radio UWB is a carrier-less (i.e., baseband) radio technology and accordingly, in this radio technique no mixer is needed. Therefore, the implementation of such a system is simple, which means that low cost transmitters/receivers can be achieved when compared to the conventional radio frequency (RF) carrier systems. The UWB technology has a history dating back to one hundred years ago when G. Marconi sent the first ever wireless transmission from the Isle of Wight to Cornwall on the British mainland using spark-gap transmitters [2]–[4]. Radio subsequently was developed to provide telephony services based on analogue techniques, and recently transited to digital telephony thanks to numerous technological advances. Through the years (1960s–1990s) the United States military developed the UWB technology that was first used for ground penetrating radar. In 1998, the Federal Communication Commissions (FCC) recognized the significance of UWB technology and initiated the regulatory review process of the technology. Consequently, in February 2002 the FCC report appeared, in which UWB technology was

H. Nikookar, R. Prasad, *Introduction to Ultra Wideband for Wireless Communications*, Signals and Communication Technology,
DOI 10.1007/978-1-4020-6633-7_2, © Springer Science+Business Media B.V. 2009

authorized for the commercial uses with different applications, operating frequency bands as well as the transmitted power spectral densities [5].

Because UWB systems operate in a very large bandwidth, they need to share the spectrum with other users as well as with the existing communication systems and consequently, interferences may occur. Besides from the interference from other users, the UWB propagation channel will cause disturbances. These important issues will be discussed in detail in the next chapters.

Here in this chapter we study the basics of ultra wideband technique for wireless communications. First in Section 2.1 the definition of UWB signals is studied. The FCC radiation mask for the UWB transmission is discussed in Section 2.2. Different UWB pulse shapes are explained in Section 2.3 and finally in Section 2.4 the major features of UWB which are remarkable for wireless communications applications are mentioned.

2.1 UWB Definition

The proposed definition by FCC for UWB transmission is: any signal, which has a fractional bandwidth (B_f) larger than 0.20, or which occupies a bandwidth greater than 500 MHz, i.e.,

$$\begin{aligned} B_f &\geq 0.2, \quad \text{or} \\ BW &> 500\,\text{MHz} \end{aligned} \tag{2.1}$$

The fractional bandwidth is defined as the ratio of signal bandwidth to the center frequency [1] and is given by

$$B_f = \frac{BW}{f_c} = \frac{(f_H - f_L)}{(f_H + f_L)/2} \tag{2.2}$$

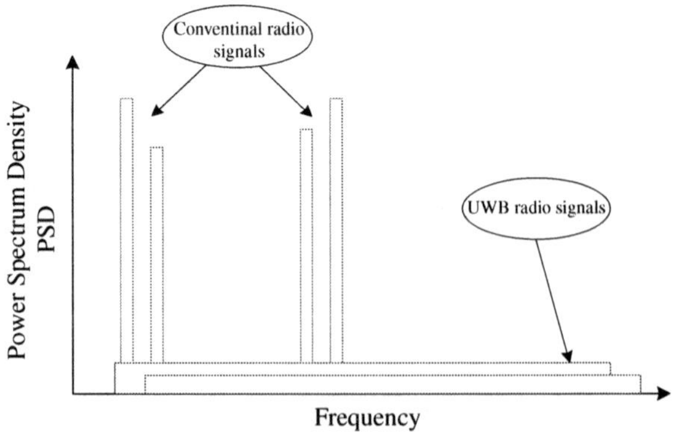

Fig. 2.1 The spectrum of the UWB signals versus conventional signals

where f_H and f_L are the highest and the lowest transmitted frequencies at the −10 dB emission point, respectively, BW is the signal bandwidth and f_c is the center frequency. As shown in Fig. 2.1 the conventional radio transmission systems (i.e. narrowband as well as wideband systems) have small fractional bandwidths when compared to the UWB signals. As an example, consider the UMTS mobile communication system which operates around 2 GHz with a bandwidth of 5 MHz. This system is often called wideband, however according to Equation (2.2), the fractional bandwidth is 0.0025, which is much smaller than 0.2 (i.e., 80 times smaller)!

2.2 FCC Mask

To avoid interference with existing communication systems, various regions of the spectrum should have different allowed power spectral densities. The Federal Communications Commission (FCC) has assigned the effective isotropic radiated power (EIRP) allowed for each frequency band [6]. EIRP is the equivalent isotropically radiated power which is the power radiated by an omni directional antenna with gain 1.The FCC mask depicts the allowed power spectral densities for specific frequencies. Figure 2.2 and Table 2.1 illustrate the FCC radiation limits for the indoor UWB communication system. The level of −41.3 dBm/MHz in the frequency range of 3.1–10.6 GHz is set to limit interference to existing communication systems, and to protect the existing radio services. This level (−41.3 dBm/MHz), is 75 nW/MHz which is in fact at the unintentional radiation level of television sets or monitors (FCC part 15 limit).

For the UWB communications the FCC has assigned two FCC masks for the indoor and outdoor UWB devices. For the indoor and outdoor UWB communications, the FCC radiation limits in the frequency range of 3.1–10.6 GHz are alike. While for the 1.61–3.1 GHz frequency range the outdoor radiation limit is

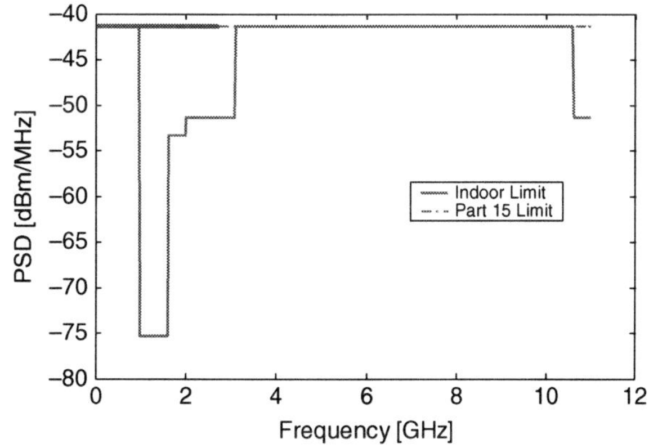

Fig. 2.2 The FCC Emission limits for the indoor UWB communications

Table 2.1 The FCC emission limits for indoor and outdoor UWB

Frequency Ranges	Indoor EIRP (dBm/MHz)	Outdoor EIRP (dBm/MHz)
960 MHz–1.61 GHz	–75.3	–75.3
1.61 GHz–1.99 GHz	–53.3	–63.3
1.99 GHz–3.1 GHz	–51.3	–61.3
3.1 GHz–10.6 GHz	–41.3	–41.3
Above 10.6 GHz	–51.3	–51.3

10 dB lower than the indoor mask. The FCC mask for the outdoor UWB communication devices are shown in Fig. 2.3 and Table 2.1. It should be noted that according to the FCC rules, the outdoor UWB communications is confined to handheld devices with no use of fixed infrastructure.

Although in this chapter we focused on the UWB spectral mask as defined by FCC, other spectrum emission masks such as European ETSI mask or Japanese mask, etc. are also available. These emission masks and the corresponding UWB regulation efforts are discussed in more detail in Chapter 10.

UWB pulses are typically narrow time pulses of sub-nanosecond or picosecond's order, as shown in the Fig. 2.4. The amplitude of the pulse should be normalized to comply with the above-discussed FCC mask.

As illustrated in Fig. 2.5, in the impulse radio UWB communications transmitted data information is modulated onto a sequence of pulses called pulse train. A processing gain of N can be achieved by putting data information on a train of N pulses. According to Fig. 2.5 the pulses are repeated after a pulse repetition interval. When pulses are sent in regular intervals, peaks will appear in the power spectral density of the transmitted UWB signal at the locations which are the multiples of the inverse of pulse repetition interval. These peaks which are also called 'comb lines' are undesirable as they easily go above the FCC limit and accordingly interfere with the other communication systems. One method to make the spectrum more noise like is by adding a small random offset to each pulse. More details on this issue can be found in Problem 2.3.

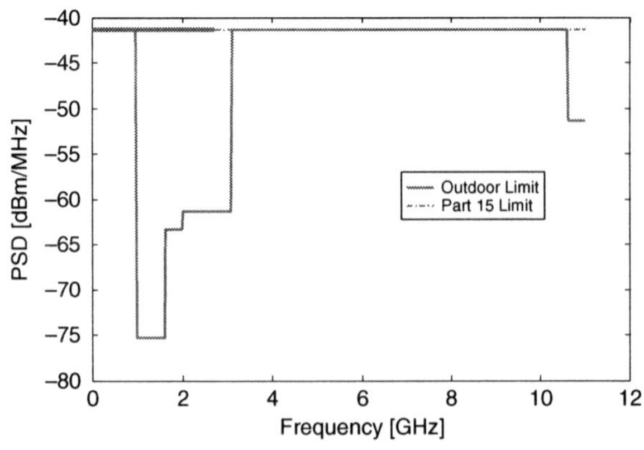

Fig. 2.3 FCC Emission limits for the outdoor UWB communications

Fig. 2.4 A typical UWB pulse shape

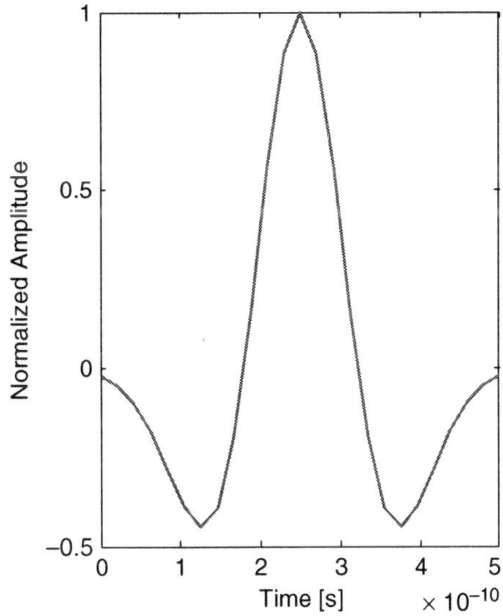

The typical signal that can be considered for the UWB transmission is the Gaussian pulse expressed as:

$$p(t) = \frac{A}{\sqrt{2\pi\sigma^2}} \mathrm{e}^{\left(-\frac{t^2}{2\sigma^2}\right)} \tag{2.3}$$

where, A and σ denote the amplitude and spread of the Gaussian pulse, respectively. The Gaussian pulses are frequently used in the UWB systems since they can be easily generated by pulse generators (when compared with the rectangular pulses with very short rise and fall time). Usually higher derivatives of Gaussian shape are more popular for the UWB transmission. This is mainly due to the DC value of the Gaussian pulse. As antennas are not efficient at DC, it is preferable to use derivatives of Gaussian shape having smaller DC components. The nth derivative of Gaussian pulse can be obtained recursively from the following expression:

$$p^{(n)}(t) = -\frac{n-1}{\sigma^2} p^{(n-2)}(t) - \frac{t}{\sigma^2} p^{(n-1)}(t) \tag{2.4}$$

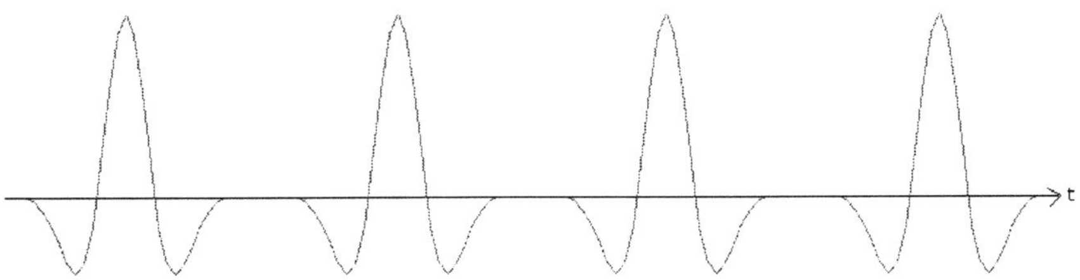

Fig. 2.5 The UWB pulse train

In Fig. 2.6 the first derivative (so called mono pulse) and second derivative (so called Doublet) of a Gaussian pulse and their spectrum are shown. In Fig. 2.7 the shape of a 7th order Gaussian pulse is shown. The power spectral density (PSD) shape of various orders of derivation of Gaussian pulse is depicted in Fig. 2.8. It is worth mentioning that due to the properties of transmit and receive antennas, which are usually modeled as differentiators the received pulse is further differentiated by the antennas.

The Fourier transform of the nth derivative of Gaussian pulse (2.4) is written as:

$$P_n(f) = A(j2\pi f)^n e^{\left(-\frac{(2\pi f\sigma)^2}{2}\right)} \tag{2.5}$$

The frequency at which the maximum value of Equation (2.5) is attained is the peak emission frequency and is given by f_M and can be found by differentiating (2.5) and setting it equal to zero. The power spectral density of the transmitted signal (PSD_t) is given by the following equation:

$$|PSD_t(f)| = A_{\max}|PSD_n(f)| = A_{\max}\frac{(2\pi f\sigma)^{2n}}{n^n}\frac{e^{\left(-(2\pi f\sigma)^2\right)}}{e^{(-n)}} \tag{2.6}$$

where $A_{\max}$ is the peak PSD that FCC permits and $|PSD_n(f)|$ is the normalized PSD of the pulse shape and is given by:

$$|PSD_n(f)| = \frac{|P_n(f)|^2}{|P_n(f_M)|^2} = \frac{(2\pi f\sigma)^{2n}}{n^n}\frac{e^{\left(-(2\pi f\sigma)^2\right)}}{e^{(-n)}} \tag{2.7}$$

In the UWB pulse shape design the aim is to obtain a pulse waveform that complies with the FCC mask as closely as possible and maximizes the bandwidth as well. For the UWB indoor systems, this results in Fig. 2.8, [7].

According to this figure, to fulfill the FCC indoor emission limits, at least the 4th derivative of the Gaussian pulse should be transmitted. While for outdoor systems, the seventh or higher are proven to satisfy FCC outdoor mask [7] as illustrated in Fig. 2.9.

There are some constraints on the UWB pulses [8]. As mentioned earlier, due to inefficiency of the transmit and receive antennas at low frequencies, the UWB wave shapes with low frequency or DC components are not popular. Another important point is the UWB channel. As will be discussed in more detail in Chapter 4, the shapes of the UWB pulses change when they pass through the channel. To elaborate further, suppose the UWB channel is modeled as a linear filter with the impulse response $h(t)$. Generally, in the transmission systems based on sinusoidal signals, if the input signal has an amplitude A and frequency f_0, the output of the channel has the sinusoidal shape with the same frequency, but its amplitude is changed (to B) and its phase is shifted (by θ). See Fig. 2.10. Unlike the sinusoidal systems, if a UWB pulse (i.e., $p(t)$) is transmitted

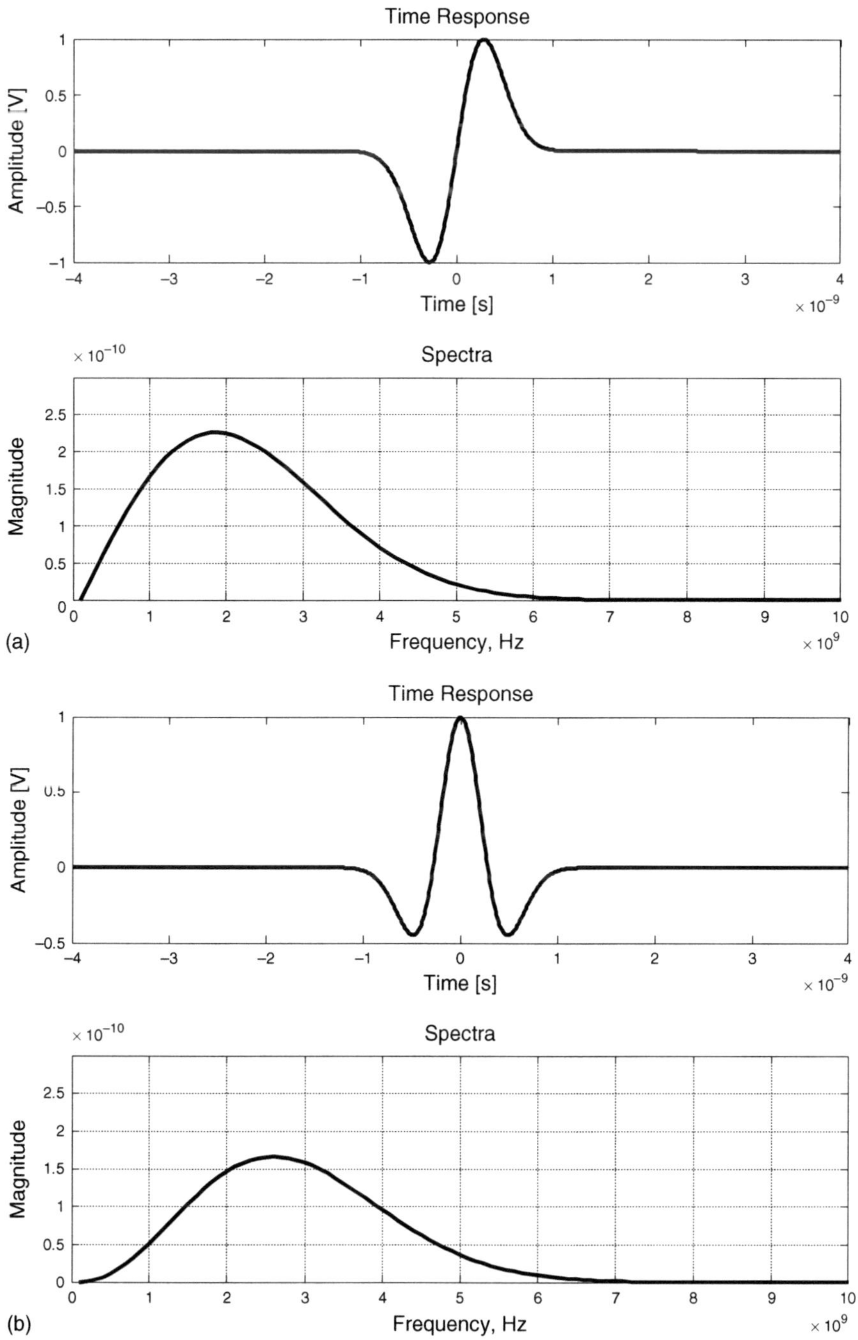

Fig. 2.6 (**a**) The Mono pulse and (**b**) Doublet and their corresponding spectra

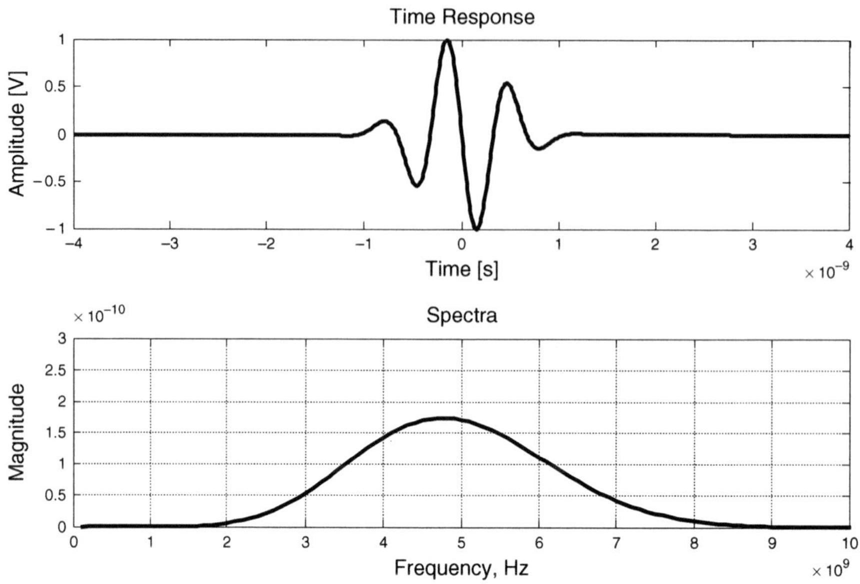

Fig. 2.7 The shape of a 7th order Gaussian pulse and its spectra

through the channel, as shown in Fig. 2.10, the shape of the out put signal (i.e., $q(t)$) is totally changed and is no longer like the input pulse shape of the channel. This is an important point to be considered when dealing with the UWB transmission and its difference with the conventional sinusoidal systems.

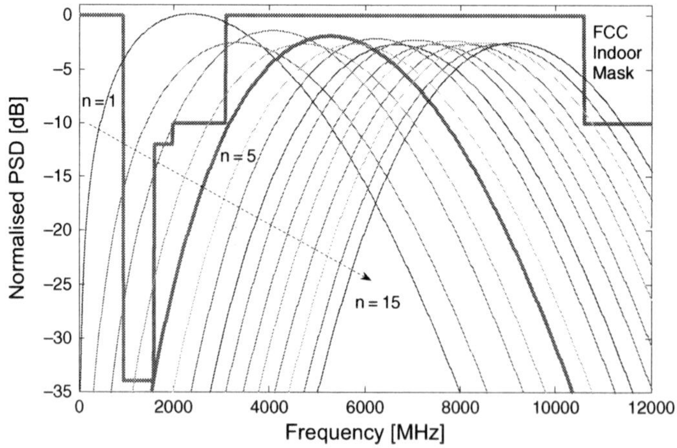

Fig. 2.8 PSD of higher order derivatives of the Gaussian pulse for UWB indoor systems

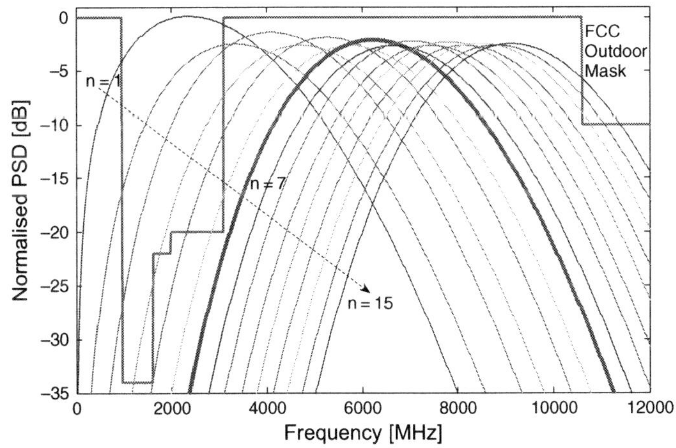

Fig. 2.9 PSD of higher order derivatives of the Gaussian pulse for UWB outdoor systems

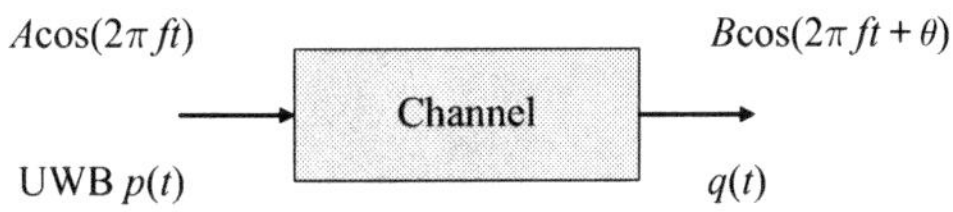

Fig. 2.10 Passing of the sinusoidal and UWB pulse shapes through a linear channel

Other waveforms can be used in UWB systems as well. The Hermite or Modified Hermite Pulses are examples of orthogonal pulses generated using Hermite polynomials that can also be used in the wireless UWB communication systems.

2.2.1 Hermite Pulses

Hermite pulses are proposed to be used in UWB wireless communications [9]. Hermite Functions appear in the field of Quantum Mechanics. The Hermite polynomial of order n is expressed as:

$$P_n(t) = (-1)^n \tau^n e^{\frac{t^2}{\tau^2}} \frac{d^n}{dt^n}\left(e^{-\frac{t^2}{\tau^2}}\right) \quad (2.8)$$

where τ is the time scaling factor, $n=0,1,2,\ldots$ and $-\infty<t<\infty$. Hermite polynomials can also be recursively obtained as [10]:

$$P_{n+1}(x) = -\frac{d}{dx}P_n(x) + 2xP_n(x) \quad (2.9)$$

where, $x = t/\tau$, with $P_0(x) = 1$, $P_1(x) = x$. The values of $P_n(x)$ are available as:

$$\begin{aligned} P_0(x) &= 1 \\ P_1(x) &= x \\ P_2(x) &= x^2 - 1 \\ P_3(x) &= x^3 - 3x \\ P_4(x) &= x^4 - 6x^2 + 3 \\ P_5(x) &= x^5 - 10x^3 + 15x \end{aligned} \tag{2.10}$$

Hermite waveforms are not orthogonal and for this reason they are modified as

$$H_n(x) = \mathrm{e}^{-x^2/2} P_n(x) \tag{2.11}$$

to become orthogonal, i.e.,

$$\int_{-\infty}^{\infty} H_n(x) H_m(x) \mathrm{d}x = 0 \qquad m \neq n \tag{2.12}$$

Orthogonality is an important issue in the wireless UWB communications as it guarantees unique demodulation of data at the UWB receiver. In Fig. 2.11 the time and frequency response of different orders (n) of the modified Hermite functions are shown. Some characteristics can be seen from this Figure. For instance the modified Hermite Pulses have almost the same duration and bandwidths for various n, and the number of zero crossings of the waveform is equal to n, [9].

2.2.2 Legandre Pulses

The Legandre polynomials have been used in many different mathematical and physical areas. Legendre polynomials are a complete orthogonal set of functions. One of various forms of the Legendre functions is expressed as:

$$P_n(t) = \frac{\tau^n}{2^n n!} \frac{\mathrm{d}^n}{\mathrm{d}t^n} \left(\frac{t^2}{\tau^2} - 1 \right)^n \tag{2.13}$$

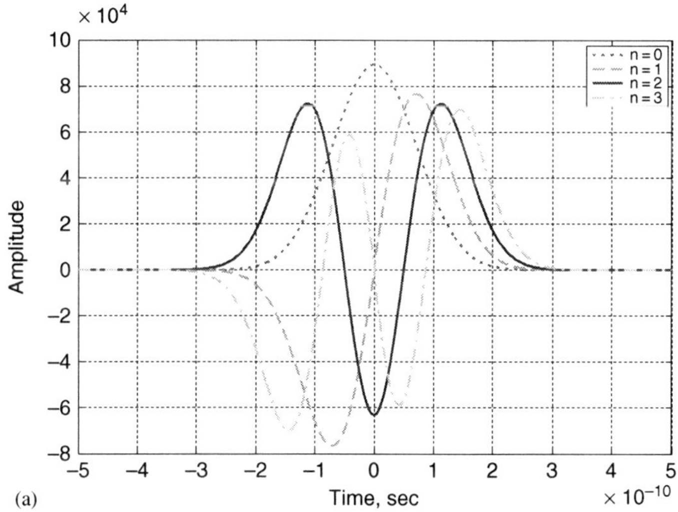

(a)

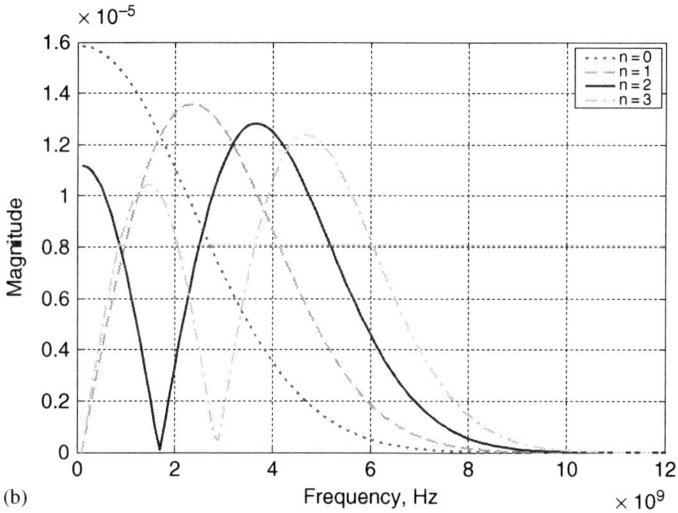

(b)

Fig. 2.11 Different orders of the modified Hermite functions, **(a)** Time shape and **(b)** frequency spectrum

where τ is the time scaling factor, $n = 0,1,2,\ldots$ and $-\infty < t < \infty$. Legandre polynomials can also be recursively obtained as [10]:

$$P_{n+1}(x) = 2xP_n(x) - P_{n-1}(x) - \frac{1}{n+1}(xP_n(x) - P_{n-1}(x)) \qquad (2.14)$$

where, $x = t/\tau$, with $P_0(x) = 1$, $P_1(x) = x$. The values of $P_n(x)$ are available as:

$$\begin{aligned} P_0(x) &= 1 \\ P_1(x) &= x \\ P_2(x) &= \frac{1}{2}(3x^2 - 1) \\ P_3(x) &= \frac{1}{2}(5x^3 - 3x) \\ P_4(x) &= \frac{1}{8}(35x^4 - 30x^2 + 3) \\ P_5(x) &= \frac{1}{8}(63x^5 - 70x^3 + 15x) \end{aligned} \tag{2.15}$$

The Legendre functions are orthogonal and it can be shown that

$$\int_{-1}^{1} P_n(x)P_m(x)\mathrm{d}x = \begin{cases} 0 & \text{if} \quad m \neq n \\ \frac{2}{2n+1} & \text{if} \quad m = n \end{cases} \tag{2.16}$$

The time shape and spectrum of different orders of the Legendre functions are illustrated in Fig. 2.12.

2.2.3 Prolate Spheroidal Functions

Another family of pulses proposed to be used in wireless UWB communications is the Prolate Spheroidal Functions (PSF). These functions are practically time and frequency limited functions, which are the solution of the following integral equation [9], [11]:

$$\int_{-T/2}^{T/2} P_n(x)\frac{\sin B(t-x)}{\pi(t-x)}\mathrm{d}x = \lambda_n P_n(t) \tag{2.17}$$

where, B is the bandwidth and T is the duration of the pulse $P_n(t)$ which is the PSF of order n. The parameter λ_n in (2.17) is the fraction of pulse energy in the interval T, i.e.,

$$\lambda_n = \frac{\int_{-T/2}^{T/2} |P_n(t)|^2\mathrm{d}t}{\int_{-\infty}^{\infty} |P_n(t)|^2\mathrm{d}t} \tag{2.18}$$

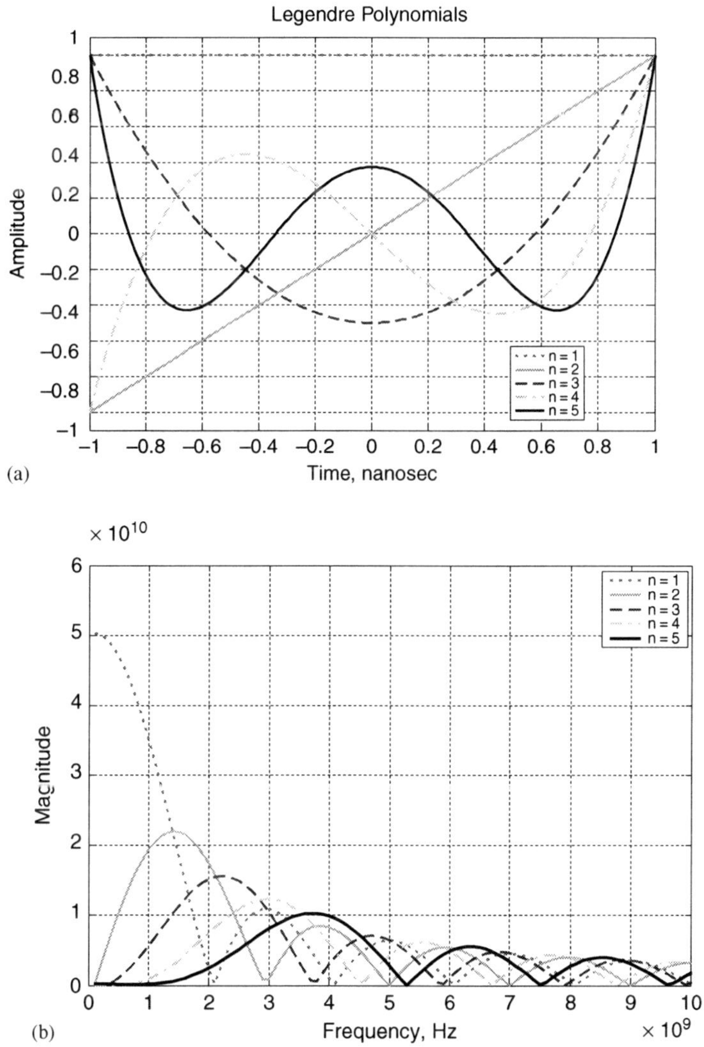

Fig. 2.12 (**a**) Time and (**b**) spectrum of different orders of the Legendre functions

The PSFs are orthogonal, i.e.,

$$\int_{-T/2}^{T/2} P_n(t)P_m(t)\mathrm{d}t = \begin{cases} 0 & \text{if} \quad m \neq n \\ \lambda_n & \text{if} \quad m = n \end{cases} \tag{2.19}$$

The time and spectrum of different orders of the PSF functions are illustrated in Fig. 2.13.

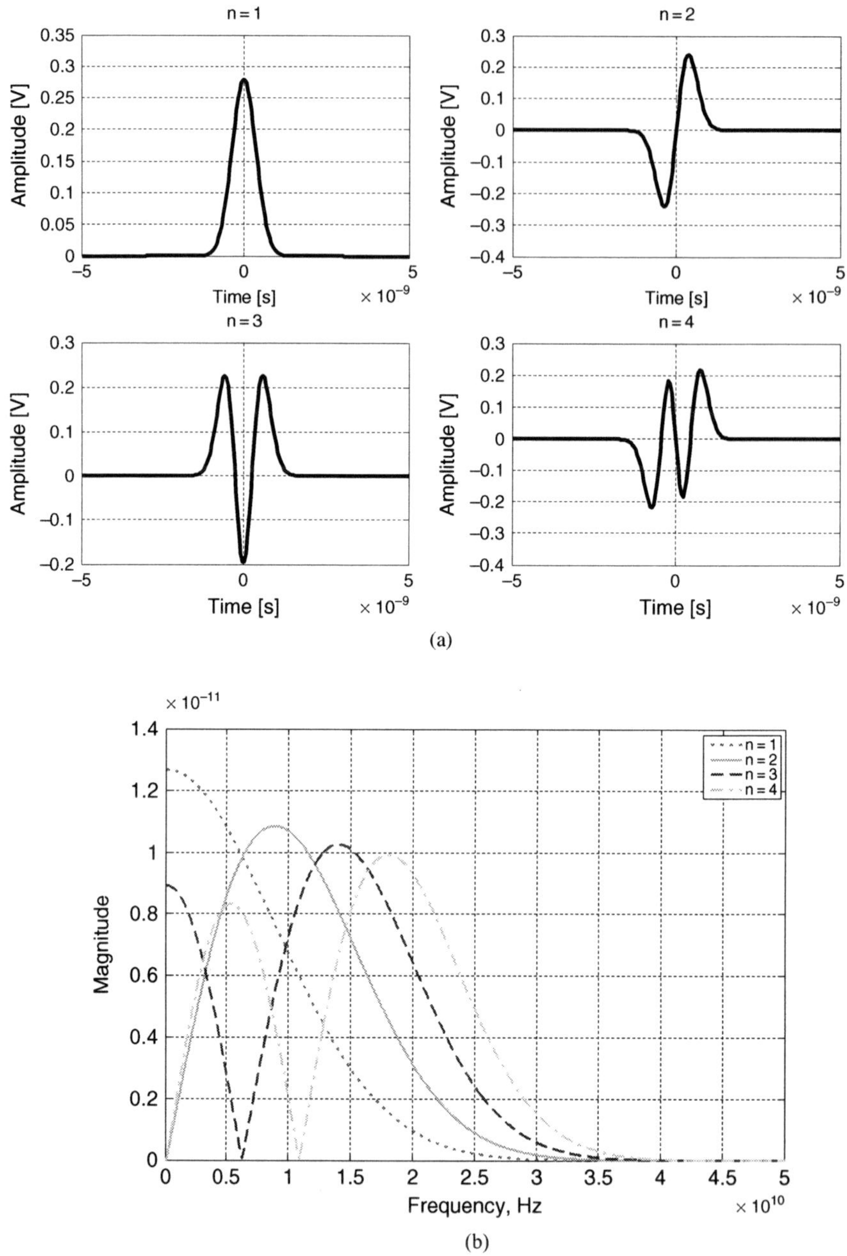

Fig. 2.13 Different orders of Prolate Spheroidal functions: (**a**) in time, (**b**) in frequency

Some properties of PSF that make them attractive for wireless UWB communications are [9]:

- The pulse duration is exactly the same for all values on n,
- The pulse bandwidth is almost the same for all values of n,
- The pulses are orthogonal,
- For large n, the pulses have a zero DC component,
- Pulse duration and bandwidth can be controlled simultaneously.

Different pulse shapes have been studied in this section. Other pulse shapes can also be designed for UWB wireless communications. Waveform design for the UWB communication is a major issue. The waveform should have a wide bandwidth and should comply with the FCC or other regulatory masks. Furthermore, the pulses should be orthogonal and if necessary should mitigate interference in the UWB band. Hence, choosing the right waveform with limitation of transmit power and efficient and dynamic spectrum management is a delicate and overwhelming task [12], [13].

2.3 UWB Features

As mentioned earlier the bandwidth of the UWB technique is huge. This very wide bandwidth means a fine time resolution. This main feature of the UWB technology provides the capability of accurate positioning which has already been used in the radar applications and is now underway in the wireless communications. The capability of communications and positioning (with precise performance), in a single technology (i.e., fusion of positioning and data capabilities in a single technology) is one of the salient features of the UWB technology.

Referring to the spectrum of the UWB signal we realize that the UWB center frequency is relatively low. This causes the UWB signal to penetrate many materials and providing a functionality that would not be present in a system of comparable bandwidth at the significantly higher center frequencies.

Besides from the high performance of the UWB technique at low cost, another major feature of this technique is the very low transmit power. This low transmit power (in the order of microwatts) causes a low level of interference to the existing systems. Moreover, as will be discussed in detail in Chapter 4, the UWB method is robust against fading. This robustness further reduces the required transmit power of this technology.

2.4 Summary

In this chapter the use of ultra wideband technique for wireless communications was motivated. Various topics related to the applicability of UWB to wireless systems including the definition of UWB signals, the FCC radiation mask for the UWB transmission, different UWB pulse shapes that can be used and the major features of UWB, which are remarkable for wireless communications applications, were studied.

Problems

Problem 2.1 Consider a Gaussian pulse shape as

$$x(t) = \frac{1}{\sigma\sqrt{2\pi}} e^{-t^2/(2\sigma^2)}$$

where σ is the time spread of the pulse. Show that the spectrum of this pulse is:

$$X(f) = e^{-\frac{(2\pi f\sigma)^2}{2}}$$

What can you conclude?

Problem 2.2 In this problem we want to study the effect of repetition of UWB pulses on the spectrum shape of the transmitted signal. Consider that data is modulated onto a train of $N = 10$ square pulses as shown in Fig. P.2.2.1. The width of each pulse is τ and pulse repetition interval (PRI) is T_0. Obtain the spectrum of the transmitted signal if $\tau/T_0 = 0.1$. What will happen to the spectral lines if instead of 10 pulses 5 pulses are transmitted and PRI is changed to $2T_0$? Keep the width of each pulse τ constant.

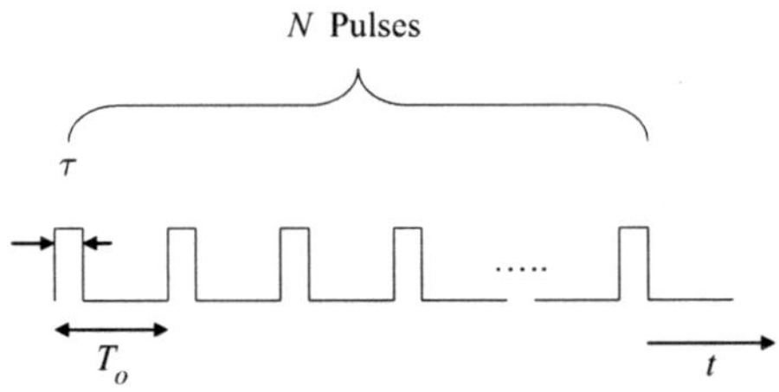

Fig. P.2.2.1 A train of UWB rectangular pulses

Problem 2.3 Consider a UWB signal which consists of a sequence of Pulse Amplitude Modulated (PAM) pulses with the following properties [14]:

- Each pulse is rectangular and of duration T_b,
- Pulses are equally likely to be $\pm$ 1,
- All pulse amplitudes are statistically independent,
- The pulses are not synchronized, i.e., the starting time T of the first pulse is equally likely to be anywhere between 0 and T_b.

Obtain the power spectral density of the UWB signal and compare it with case where there is no uncertainty in the starting time of first pulse.

Problem 2.4 The impulse response of a UWB channel can be represented as

$$h(t) = e^{-\alpha t} u(t)$$

where α^{-1} is the decay time constant of the impulse response profile and $u(.)$ is the unit step function.

(a) Calculate the output of the channel if a sinusoidal signal $p(t) = A \cos(2\pi f_0 t)$ is passed through the channel (where, A is the amplitude and f_0 is the frequency of the sinusoidal signal).
(b) Repeat part (a) if a periodic pulse $p(t)$ with period of T_0, with the following shape passes through the channel:

$$p(t) = \begin{cases} 1 & \text{if} \quad |t| \leq \tau/2 \\ 0 & \text{otherwise} \end{cases}$$

For the calculations consider $\tau = 0.05T_0$ and $\alpha = 1/T_0$. Compare the result with part (a).

Problem 2.5 Using the definition of fractional bandwidth calculate the fractional bandwidth for the WLAN 802.11a and Bluetooth Wideband systems and compare them to the fractional bandwidth of the UWB.

Problem 2.6 Calculate the fractional bandwidth of a Doublet UWB pulse derived from a Gaussian pulse with $\sigma = 0.75$ nsec. around center frequency of 2.4 GHz. If this pulse shape is modulated around 8 GHz what will be the fractional bandwidth?

Problem 2.7 Show that for the Gaussian pulse, further derivatives make more zero crossing of the waveform. Also show that this will further decrease the fractional bandwidth of the signal.

References

1. Online reference:http://www.fcc.gov/oet/info/rules.
2. K. Siwiak, "Ultra-wideband radio: A new pan and positioning technology," IEEE Vehicular Technology Society News, February 2002, pp. 4–9.
3. K. Siwiak, P. Withington and S. Phelan, "Ultra-wideband radio: The emergency of an important new technology," IEEE Vehicular Technology Conference, VTC-Spring, vol. 2, no. 53, pp. 1169–1172, 2001.
4. K. Siwiak, "Ultra-wideband radio: Introducing a new technology," IEEE Vehicular Technology Conference, VTC-Spring, vol. 2, no. 53, 2001, pp. 1088–1093.
5. FCC, "Revision of Part 15 the Commission's Rules Regarding Ultra-wideband Transmission Systems," ET Docket, 2002.
6. S. Hongson et al., "On the spectral and power requirements for UWB transmission," ICC 2003, vol. 1, May 2003, pp. 738–742.

7. H. Sheng, "Transceiver design and system optimization for Ultra-Wideband communications," Ph.D. Dissertation, New Jersey Institute of Technology, Newark, May 2005.
8. J.R. Davis et al., "Some physical constraints on the use of carrier free waveforms in the radio wave transmission systems," Proc. IEEE, June 1979, pp. 884–890.
9. M. Ghavami, L.B. Michael and R. Kohno, Ultra Wideband Signals and Systems in Communication Engineering, Wiley, 2004.
10. G.B. Arfken and H.J. Weber, Mathematical Methods for Physicists, Elsevier Academic Press, MA, USA.
11. H.L. Van Trees, Detection, Estimation and Modulation Theory, Pt-1, John Wiley, NY, 1968.
12. H. Nikookar, Wavelet Radio: Smart, Adaptive and Reconfigurable Wireless Systems based on Wavelets, Tutorial, WPMC2007, 3–6 December 2007, Jaipur, India.
13. H. Nikookar, Wavelets for Wireless Communication, Tutorial, European Conference on Wireless Technology, Munich, Germany, October 2007.
14. K.S. Shanmugham, Digital and Analog Communication Systems, John Wiley, 1985, NY, USA.

Chapter 3
UWB Antennas

Contents

Over the past few years Ultra Wideband communication has received much attention. Research and development on UWB communications is gaining momentum mainly because of communication and ranging capabilities of this technology as well as the recognition of significance of UWB technology by the FCC and other regulatory bodies. The major step in the development of UWB technology for wireless communications is the antenna. Like all wireless devices, the antenna is an essential part of UWB communication systems. Due to huge bandwidth of the UWB system, antenna poses a remarkable challenge to the UWB technology. In the recent years researchers, engineers and scientist have tried hard to solve UWB antenna problem in different ways and now, six years after the FCC authorized commercial UWB systems, the UWB products are entering the market with a large scale of commercialization.

Generally antennas are elements that radiate the electromagnetic energy of a transmission line to the free space. Antennas are in fact transition devices (transducers) between guided wave and free space (and vice versa), [1]. They can be considered as impedance transformers, coupling between an input or line impedance and the impedance of free space [2]. For the case of the UWB this impedance transformation of antenna is more important. This is due to huge bandwidth of UWB system. As an initial approach to the UWB antennas we can start from a dipole and consequently consider multi-narrowband antennas which are optimized to work in the entire UWB band. This idea is shown in Fig. 3.1 together with the antenna's corresponding dispersive waveform. The large scale components of this log-periodic antenna radiate the low frequency components and the smaller scale components of the antenna radiate high

H. Nikookar, R. Prasad, *Introduction to Ultra Wideband for Wireless Communications*, Signals and Communication Technology,
DOI 10.1007/978-1-4020-6633-7_3, © Springer Science+Business Media B.V. 2009

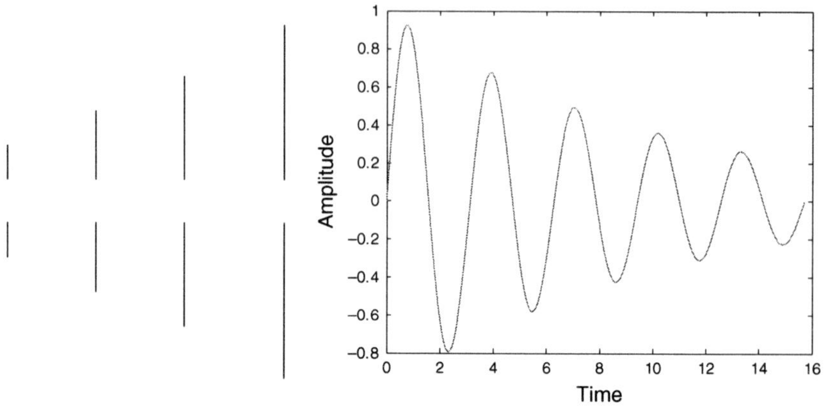

Fig. 3.1 A log-periodic antenna (*left*) which has a dispersive waveform (*right*)

frequency components. For the UWB communications the dispersive behavior of the antenna waveform is not popular. Another disadvantage of this antenna is at different azimuth angles around the antenna the waveform varies, which is again unpopular for wireless communication applications.

There are different types of UWB antennas. As described in more detail in [3] they are categorized into the following classes according to form and function:

- Frequency dependent antennas: The log-periodic antenna is an example of this type of antennas where the smaller scale geometry of antenna contributes to higher frequencies and the larger scale part contributes to the lower frequencies.
- Small-element antennas: These are small, omni-directional antennas for commercial applications. Examples of this type of antennas are bow-tie or diamond dipole antennas.
- Horn antennas: Horn antennas are electromagnetic funnels that concentrate energy in a specific direction. These antennas have large gains and narrow beams. The Horn antennas are bulkier than small-element antennas.
- Reflector antenna: These antennas are high gain antennas that radiate energy in a particular direction. They are relatively large but easy to adjust by manipulating the antenna feed. Hertz's parabolic cylinder reflector is an example of this type of antennas.

Reference [3] is an excellent book with a comprehensive treatment of UWB antennas. The aim of this chapter is to provide a brief introduction to the application of UWB antennas especially for wireless communications. To this end, first in Section 3.1 we start with explaining the requirements of the UWB antennas when they come to wireless communication applications. The radiation mechanism of the UWB antennas is explained in Section 3.2. Since design of a UWB communication system requires a good understanding of the link budget for determining the coverage of the communication system, in

Section 3.3 a link budget for UWB system, taking into account the characteristics of the UWB antennas, is provided. Furthermore, as UWB is a proper technology for the short range high-speed communications, and in short range scenarios the UWB transmit and receive antennas are very close to each other and the far-field condition, assumed in most of the link budget models, may not be satisfied, in Section 3.4 a short range analysis of UWB antennas is discussed.

3.1 Antenna Requirements

The major challenge in the transmission of narrowband pulses is the antenna. Unlike the narrowband sinusoidal regime, the radiation of large bandwidth, non-sinusoidal waveforms is an active field of research in the antenna domain. In this section we study the major requirement of antennas for UWB wireless communications.

As discussed in Chapter 2, the UWB bandwidth spans from 3.1 to 10.6 GHz. This huge bandwidth poses challenges on the antennas of the UWB system. The antenna's characteristic should be flat in this frequency range. The UWB antennas for wireless communications applications should have omni-directional radiation pattern. They should radiate the pulse with minimal distortion and minimal late-time ringing [4]. They should also be integrated with the generator on a chip. UWB antennas should be mounted on a dielectric substrate which will also serve as a protective mechanical shield and radiate through it. Other major constraint is the FCC mask. In fact the antenna designer and the RF engineer should cooperate to ensure the overall UWB communication device to meet a desired spectral mask [2]. Other requirement of UWB antennas for wireless communications emanates from the spectrum sharing of the UWB and the existence of other narrowband services. The wireless communications UWB antennas should be able to be modified by introducing notches or filtering to some frequencies occupied by the narrow and wideband services. Another requirement is the integration of UWB antenna with the UWB device. UWB antenna for wireless communications should be an integral part of the system and not a stand-alone element. This is an important issue in the successful implementation of UWB technology for wireless communication applications.

3.2 Radiation Mechanism of the UWB Antennas

Wireless systems based on impulse radio use a new temporal dimension not classically considered in antenna theory and well-known frequency domain parameters are not sufficient for the characterization of transient radiation behavior. In the UWB measurements, the pulse generator is connected to the UWB antenna. As the antenna is often not perfectly matched at low frequencies, when the transmitted pulse arrives at the transmit antenna, not all the energy is transmitted. This consequently means that a part of the signal is reflected. This reflected pulse travels back through the cable to the generator. Because the generator is short circuited after firing the pulse, this reflected pulse is then reflected back to the antenna and is partially transmitted into the channel and the process repeats. Therefore, the pulse bounces between the UWB generator and UWB transmit antenna. This phenomena is known as "multiple signal reflection" in the UWB antenna measurements [5]. There are several solutions to this problem. A standard solution is to use a wideband resistive attenuator between generator and transmit antenna to attenuate reflections at the price of range reduction due to loss of transmit power. Another solution to this problem is the use of an isolator or circulator. However, these elements should operate over a large bandwidth. In the measurements, a practical solution to this problem is to use a longer cable between the UWB generator and transmit antenna. The length of the cable determines the total time window that can be measured in the UWB measurements.

3.3 Link Budget for UWB System Taking into Account the UWB Antennas

Design of a UWB communication system requires a good understanding of the link budget for determining the coverage of the system. Indeed several results have been reported on the UWB channel measurements and modeling and accordingly, models for UWB path loss have been proposed [6]–[9]. However, none of these references has investigated the link budget of UWB transmission when taking into account the characteristic of UWB antennas. In this section we study a link budget model for a UWB system considering the characteristics of the transmit and receive antennas as well as the generator and pulse shape of the UWB system.

The electric field radiated by the transmit antenna E^{rad}can be expressed as [10]

$$E^{\mathrm{rad}}(f) = \frac{jV_g(f)\eta_0}{Z_g(f) + Z_A^T(f)} F_0 \frac{\mathrm{e}^{-j\beta R}}{4\pi R} \tag{3.1}$$

Where V_g is the generator output voltage and Z_g and Z_A^T are the impedance of the generator, and the transmit antenna, respectively. The following parameters are used: f is the frequency, $\beta = 2\pi/\lambda$ is the free space wave number, where λ is the wavelength, $\eta_0 = 120\pi$ is the free space impedance, F_0 is the antenna field factor and R is the distance. If the transmit and receive antennas are assumed to be identical, then the field factor can be expressed as [11]

$$F_0 = \beta_0 H_e \tag{3.2}$$

Where H_e is the effective height of the antenna. The open circuit voltage induced at the receive antenna output is proportional to the incident electric field E^{inc} and the effective height of the antenna. Assuming $E^{\mathrm{inc}} = E^{\mathrm{rad}}$, the voltage at the load, i.e., V_L can be expressed as (see Fig. 3.2)

$$V_L(f) = I_L(f)Z_L(f) = E^{\mathrm{inc}} H_e \frac{Z_L}{Z_A^R + Z_L(f)} \tag{3.3}$$

Where I_L is the current through the load, Z_A^R and Z_L are the impedance of the receive antenna and load, respectively. Inserting (3.1) and (3.2) in (3.3) yields:

$$V_L(f) = j\beta H_e^2 \eta_0 \frac{\mathrm{e}^{-j\beta R}}{4\pi R} \frac{Z_L(f)}{\left(Z_A^R + Z_L(f)\right)\left(Z_A^T + Z_g(f)\right)} V_g(f) \tag{3.4}$$

According to (3.4), the voltage at the load can be determined if the parameter H_e is known as other parameters are known from the design. From (3.4) the received power spectral density can be found:

$$P_r(f) = P_t(f)\beta^2 H_e^4 \eta_0^2 \frac{1}{(4\pi R)^2} \frac{|Z_L(f)Z_g(f)|}{\left|Z_A^R + Z_L(f)\right|^2 \left|Z_A^T + Z_g(f)\right|^2} \tag{3.5}$$

Where

$$P_t(f) = \frac{1}{2} \frac{|V_g(f)|^2}{|Z_g(f)|} \tag{3.6}$$

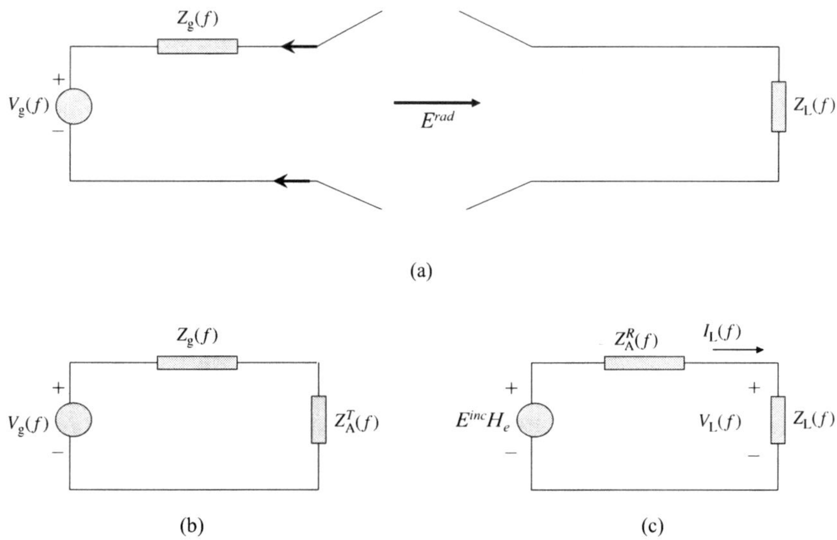

Fig. 3.2 Illustration of (**a**) antenna system, (**b**) transmit equivalent circuit and (**c**) receive equivalent circuit

and

$$P_r(f) = \frac{1}{2}\frac{|V_L(f)|^2}{|Z_L(f)|} \quad (3.7)$$

are the transmit and receive power spectral densities, respectively. The effective antenna height can be expressed in terms of the antenna gain and its input impedance as:

$$H_e^4 = \frac{|Z_A^R + Z_L(f)|^2 |Z_A^T + Z_g(f)|^2}{|Z_L(f) Z_g(f)|} \frac{\lambda^4}{4\pi^2 \eta_0^2} G_t(f) G_r(f) \quad (3.8)$$

Where $G_t(f)$ and $G_r(f)$ are the transmit and receive antenna gain, respectively. Substituting (3.8) in (3.4) we get the following expression for the link budget based on signal amplitudes:

$$V_L(f) = \left(\frac{c}{4\pi} \frac{\sqrt{G_t(f) G_r(f)}}{f} |V_g(f)| \right) \frac{1}{R} \quad (3.9)$$

Where c is the speed of light. According to (3.9) the link budget expression is based on the received voltage rather than the received power. Such an approach is more appropriate for a system working in the time domain. Calculation of link budget based on peak voltage can avoid the noise level problem which is caused by averaging the signal over a small time window [12].

Using the above-mentioned analysis, for a transmitted pulse of 200 ps half-pulse width (i.e., pulse width at half of the maximum amplitude), providing a spectrum of about 6 GHz, and 2 set of omni-directional bi-conical UWB antennas (antenna set 1 with diameter of 16 cm and set 2 with diameter of 7.5 cm) and with antenna height of 1.5 m, the maximum received voltage as a function of separation distance between transmitter and receiver is shown in Fig. 3.3, [12]. From this figure it can be observed that the received voltage decreases inversely with the separation distance between antennas. The simulation results based on (3.9) show a good match with the measurements as discussed in detail in [12]. Moreover, the received voltage with antenna set 1 is much higher than the case when antenna set 2 is used. This is because antenna set 1 radiates more energy than antenna set 2 at low frequencies where the major part of the pulse energy is concentrated.

From (3.9) the effect of generator waveform on the link budget is also clear. Figures 3.4 and 3.5 show the rectangular, Gaussian and mono cycle pulse in time and frequency domain, respectively. The results of the above discussed analysis of link budget for the antenna set 2 are illustrated in Fig. 3.6. From this

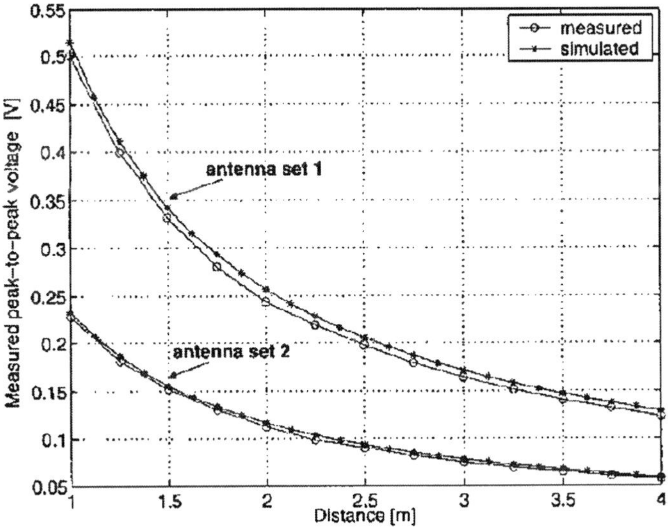

Fig. 3.3 Received peak-to-peak voltage as a function of distance using 2 sets of biconical UWB antennas, [12]

figure it is obvious that the received voltage for the monocycle pulse is higher than the rectangular and the Gaussian pulses. This is mainly because, unlike other considered waveforms, the monocycle pulse has almost no DC component and the main energy occurs at higher frequencies where the antennas perform better.

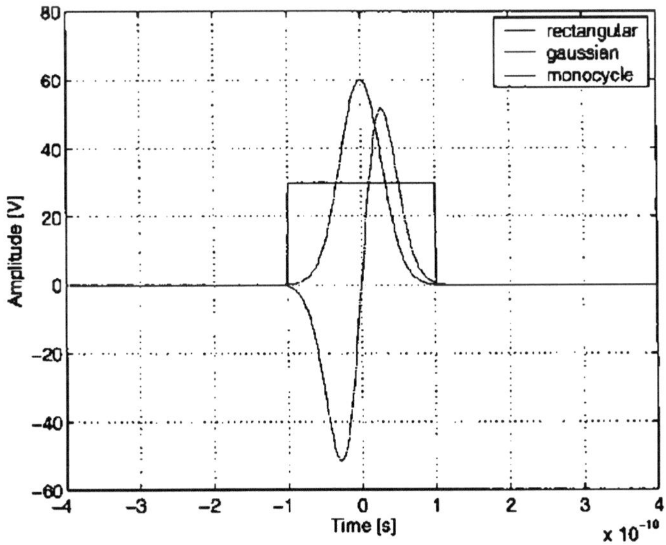

Fig. 3.4 Different shapes of the generator output with equal total energy, [12]

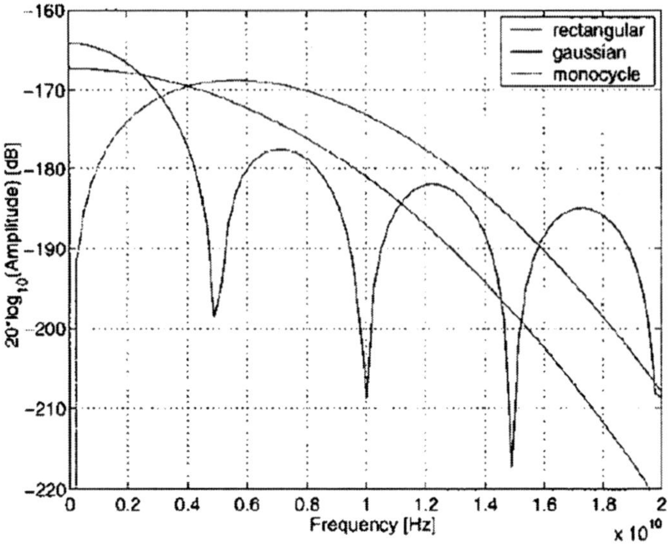

Fig. 3.5 Spectra of the pulses, [12]

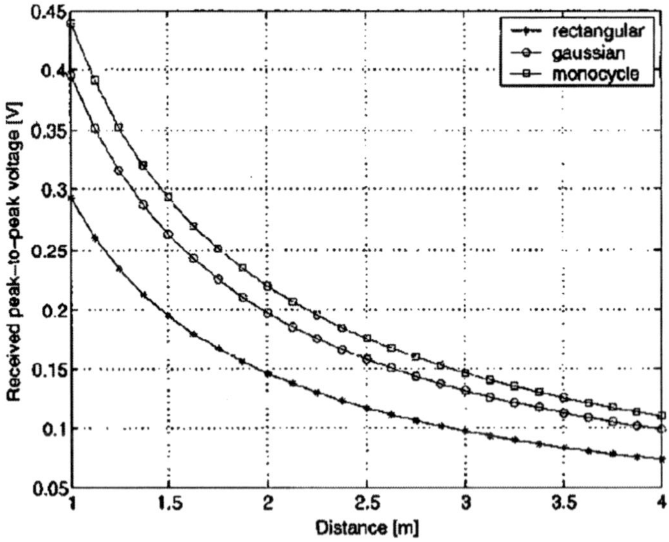

Fig. 3.6 Receive peak-to-peak voltage as a function of distance between transmitter and receiver using different pulse shapes, [12]

3.4 Short Range Analysis of UWB Antennas

UWB is a proper technology for short range high speed communications. Major application area is the Wireless Personal Area Networks (WPAN) or Wireless Body Area Networks (WBAN). In these applications the transmit and receive antennas are very close to each other and far-field condition, assumed in most of the link budget models, may not be satisfied. In the far-field, the plane wave propagation and a constant ratio between electric and magnetic field given by the wave impedance, is assumed. However, in the near field region, different effects such as reactive fields, phase errors and even re-radiation between the antennas may become important.

Different near field criteria have been established based on different effects that can appear in this region [10]. In this section these effects, among others, are analyzed with respect to the received power over a large frequency band.

3.4.1 Phase Error

The near field criterion associated to phase error is restricted to the fact that an incident spherical wave differs from a plane wave in a given fraction of the wavelength. When transmit and receive antennas are close to each other, the rays of the spherical wave produced by transmit antenna reach different parts of the receive antenna aperture with different phases. This affects the receive power as the rays do not sum coherently at the receive antenna. The phase of each ray is a function of distance between the antennas, the frequency and the antenna dimension. This effect causes variations in the received power as a function of distance between the antennas and the frequency and may cause distortion in the received signal. For the sake of simplicity, and without loss of generality, we consider a dipole antenna with full matching. We analyze the case of two thin diploes (i.e., transmit and receive) as depicted in Fig. 3.7, [13]. The voltage V_{rx} at the terminals of the receive dipole can be expressed as [10]:

$$V_{\mathrm{rx}} \infty \frac{1}{I_{\mathrm{rx}_i}} \int_{-l/2}^{l/2} E_{\mathrm{rz}}(z_{\mathrm{rx}}) I_{\mathrm{rx}}(z_{\mathrm{rx}}) \mathrm{d}z_{\mathrm{rx}} \tag{3.10}$$

Where I_{rxi} is the current at the feed point of the receive antenna, I_{rx} is the current illumination of the receive dipole, z_{rx} and z_{tx} are positions along the receive and transmit dipole, respectively, and l is the length of the dipole. The electric field along the receive dipole is given by:

$$E_{\mathrm{rz}}(z_{\mathrm{rx}}) = \int_{-l/2}^{l/2} E_{\theta}(z_{\mathrm{tx}}, z_{\mathrm{rx}}) \cos \alpha(z_{\mathrm{tx}}, z_{\mathrm{rx}}) \mathrm{d}z_{\mathrm{tx}} + \int_{-l/2}^{l/2} E_{r}(z_{\mathrm{tx}}, z_{\mathrm{rx}}) \cos\left[\frac{\pi}{2} - \alpha(z_{\mathrm{tx}}, z_{\mathrm{rx}})\right] \mathrm{d}z_{\mathrm{tx}} \tag{3.11}$$

Using the equations of the infinitesimal dipole given in [10]

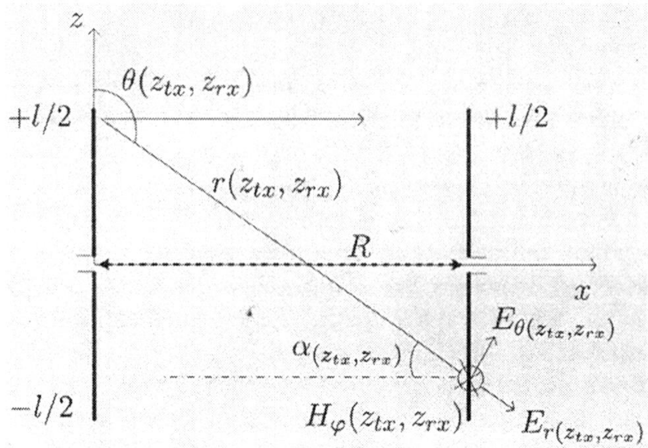

Fig. 3.7 Phase error between two dipoles

$$E_r = \eta \frac{I_0 l \cos\theta}{2\pi r^2}\left(1 + \frac{1}{jkr}\right)\mathrm{e}^{-jkr} \tag{3.12}$$

$$E_\theta = j\eta \frac{I_0 l \sin\theta}{4\pi r}\left(1 + \frac{1}{jkr} - \frac{1}{(kr)^2}\right)\mathrm{e}^{-jkr} \tag{3.13}$$

The voltage in the receive antenna can be expressed as:

$$V_{\mathrm{rx}} \infty \frac{1}{I_{\mathrm{rx}_i}} \int\limits_{-l/2}^{l/2} \left(\begin{array}{c} \int_{-l/2}^{l/2} j\eta \frac{kI(z_{\mathrm{tx}}) l \sin\theta(z_{\mathrm{tx}}, z_{\mathrm{rx}})}{4\pi r(z_{\mathrm{tx}}, z_{\mathrm{rx}})} \left(1 + \frac{1}{jkr(z_{\mathrm{tx}}, z_{\mathrm{rx}})} - \frac{1}{(kr(z_{\mathrm{tx}}, z_{\mathrm{rx}}))^2}\right)^{-jkr(z_{\mathrm{tx}}, z_{\mathrm{rx}}) \cos\alpha(z_{\mathrm{tx}}, z_{\mathrm{rx}})} \mathrm{d}z_{\mathrm{tx}} + \\ \int_{-l/2}^{l/2} \eta \frac{I(z_{\mathrm{tx}}) l \cos\theta(z_{\mathrm{tx}}, z_{\mathrm{rx}})}{2\pi r(z_{\mathrm{tx}}, z_{\mathrm{rx}})^2} \left(1 + \frac{1}{jkr(z_{\mathrm{tx}}, z_{\mathrm{rx}})}\right) e^{-jkr(z_{\mathrm{tx}}, z_{\mathrm{rx}}) \cos[\frac{\pi}{2} - \cos\alpha(z_{\mathrm{tx}}, z_{\mathrm{rx}})]} \mathrm{d}z_{tx} \end{array} \right)$$
$$I_{\mathrm{rx}}(z_{\mathrm{rx}}) \mathrm{d}z_{\mathrm{rx}} \tag{3.14}$$

Where k is the wave number $(2\pi/\lambda)$, I_0 is the excitation current, η is the free space impedance $(120\pi\Omega)$, θ is the elevation angle with respect to the axis of the dipole assumed along to zenith, φ is the azimuth angle with respect to the x-axis in a rectangular coordinate system with z-axis along the zenith, and r is the distance from the dipole.

Equation (3.14) quantitatively shows the near field effects on the received signal voltage. Calculation of received power at very short distances shows that the received power is less than expected for the far-field [13]. The main reason is that different rays picked up by the antenna are not in phase and therefore the total received power is less. It should be mentioned that for the UWB antennas in short range wireless communication scenarios, the revived power based on (3.14) should be averaged over the entire UWB band. This means that the results of different dipoles at different center frequencies should be combined. The above-explained analysis can be repeated in a similar manner for the non-dispersive UWB antennas (such as diamond antenna*). More details can be found in [13]–[14].

3.4.2 Antenna Mismatch

The electric and magnetic fields for distances within the near-field region have a quite different behavior. They consist of radiating components and receive or storage components. For radiating fields, the source is not affected by the observer. Once the radiating field leaves the antenna, it is gone for ever, and the source is not affected if the energy is absorbed or not. For reactive or stored fields the effect is different. Any time that an observer extracts or diverts part of the reactive field energy, it will cause a reaction in the source circuit which might change the input impedance of the antenna [14]. For distances below $\lambda/2\pi$ the reactive fields predominate. The coupling of the reactive static field to the receive antenna can be modeled as an electric or magnetic coupling and it will influence the transmit antenna by changing its input impedance. This coupling will vary with the frequency. For UWB antennas the changes in the input impedance of the transmit antenna may cause a mismatch for one or several frequencies, and a distortion in the frequency response of the antennas. This effect is analyzed in [14] by using a diamond antenna with a second identical antenna placed at different distances and different frequency bands. Results of this reference show that only for distances of 1 cm and less a small antenna mismatch occurs. It is also concluded that the mismatching is more important for larger UWB antennas than for smaller antennas.

3.4.3 Re-radiation Between Antennas

The voltage induced in a receive antenna due to electromagnetic field generated by a transmit antenna produces a current through both the antenna impedance and the load impedance. Part of the power picked up by the antenna will be absorbed by the load impedance and a part of power will be dissipated in the antenna impedance. This might produce a re-radiation from the receive antenna that can be picked up by the transmit antenna and re-radiated again producing interference in the receive antenna. The amount of power re-radiated by the receive antenna depends on the antenna mismatch and the physical antenna structure. For a given distance the re-radiation between the antennas may add constructively or destructively depending on the frequency and the phase shift introduced by both antennas. This effect causes a variation of the received power as a function of frequency. This means that the channel frequency response may change for different distances, not only in level, but also in the

* A diamond antenna is a planar antenna formed by two isosceles triangles. The width and the height of the triangle is about $\lambda/4$ of the center frequency.

shape. Simulation results of the UWB diamond antenna in [14] indicate that at distances of 2–3 cm the re-radiation effect between antennas are negligible. Furthermore, this effect is smaller for smaller UWB antennas.

3.5 Summary

The design and development of antennas for UWB wireless communication is a key research area. The huge bandwidth of UWB systems poses unique research challenges which have to be dexterously addressed. In this chapter a brief introduction to the application of UWB antennas especially for wireless communications was provided.

Problems

Problem 3.1 Consider an AM broadcast antenna working in the frequency range of 530-–1800 KHz. Calculate the fractional bandwidth of the signal. Can we consider the antenna as an UWB antenna? (Hint: note that the bandwidth of AM signal is 10 KHz.)

Problem 3.2 As will be discussed in more detail in Chapter 7, Impulse radio and Multi-band OFDM are two competent technologies for the UWB communications. Which one of these technologies is more vulnerable to the dispersion of the UWB antenna?

Problem 3.3 What are the major differences between UWB antennas for the radar and wireless communication applications?

Problem 3.4 Consider a linear array of N narrowband antennas which are equally spaced at a distance d. As seen in Fig. P.3.4.1, a plane wave impinges the array at angle q with respect to array normal. Each antenna output is weighted by a complex weight V_n. By adding all antenna outputs obtain the array factor (AF) as:

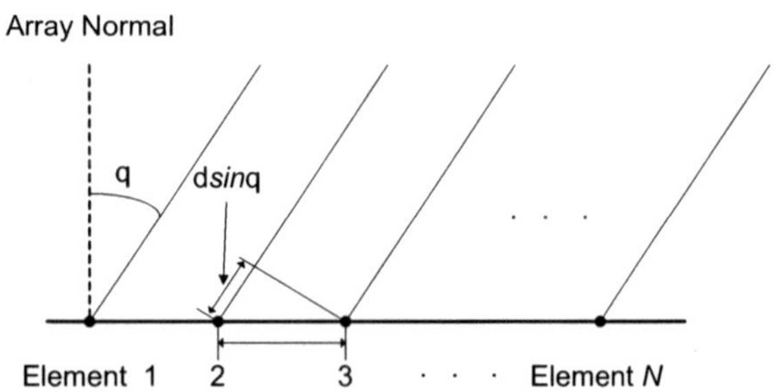

Fig. P.3.4.1 A uniformly spaced linear array

$$\mathrm{AF}(q) = \sum_{n=1}^{N} V_n \mathrm{e}^{j\frac{2\pi}{\lambda} nd \sin \mathrm{q}}$$

where λ is the wavelength.

Sketch the *AF* for an array of 8 elements.

If the array elements are UWB antennas which are modeled by differentiators, what will be the array factor? Sketch the array factor for $N=8$ and compare the result with the narrowband antennas case.

Problem 3.5 In Section 3.2 we studied the radiation mechanism of the UWB antennas and the multiple reflection problems. Does a longer cable between the UWB generator and transmit antenna eliminate the problem?

References

1. J. Kraus, Antennas, 2nd edition, McGraw Hill, NY, 1988.
2. F. Nekoogar, Ultra-Wideband Communications, Fundamentals and Applications, Prentice Hall, NJ, 2006.
3. H. Schantz, The Art and Science of Ultrawideband Antennas, Artech House, Boston, 2005.
4. A. Yarovy and K. Palmer, "Ultra-wideband Antennas: Foundations," Workshop, European Conference on Wireless Technology, Paris 2005.
5. Z. Irahhauten, G. Janssen, H. Nikookar, A. Yarovoy and L.P. Ligthart, "UWB Channel measurements and results for office and industrial environments," IEEE Proc. 2006 International Conference on Ultra-Wideband, MA, September 2006, pp. 225–230, ISBN:1-4244-0102-X.
6. D. Cassioli, M.Z. Win and A.R. Molisch, "A statistical model for the UWB indoor channel," IEEE Vehicular Technology Conference Spring, vol. 2, no. 53, 2001, pp. 1159–1163.
7. J. Keignart and N. Daniele, "Subnanosecond UWB channel sounding in frequency and temporal domain," IEEE Conference on UWB Systems and Technologies, 2002, pp. 25–30.
8. S. Yano, "Investigating the UWB indoor wireless channel," IEEE Vehicular Technology Conference Spring, vol. 3, no. 55, 2002, pp. 1200–1204.
9. K. Siwiak and A. Petroff, "A path link model for UWB pulse transmission," IEEE Vehicular Technology Conference Spring, vol. 2, no. 53, 2001, pp. 1173–1175.
10. C.A. Balanis, Antenna Theory: Analysis and Design, John Wiley, NY, 1997.
11. E.K. Muller, Time-Domain Measurements in Electromagnetics, Van Nostrand Reinhold Company, NY, 1986.
12. Z. Irahhauten, A. Yarovoy, H. Nikookar, G. Janssen and L. Ligthart, "The effect of antenna and pulse waveform on the link budget of ultra wide band impulse radio transmission," European Conference on Wireless Technology, October 2004, pp. 261–264, Amsterdam.
13. Z. Irahhauten, J. Dacuna, G. Janssen and H. Nikookar, "A link budget model for UWB-WPAN applications," European Conference Wireless Technology, September 2006, Manchester, UK, pp. 95–95–98.
14. Z. Irahhauten et al., "Link budget analysis and modeling for short-range UWB channels," Submitted to IEEE Transactions on Antenna Propagation, December 2007.

Chapter 4
Ultra Wide Band Wireless Channels

Contents

Analysis and design of UWB wireless communication systems are based on the knowledge of the propagation characteristics of UWB radio channel. Radio propagation mechanisms in the mobile and indoor environments are complex. UWB propagated radio signals undergo attenuation by objects. Meanwhile, because of the reflection, refraction and scattering of the transmitted wave, the signal arrives in the receiver through different paths having different amplitudes resulting in the delayed and attenuated echoes of each transmitted pulse. Because of the huge bandwidth, and unlike the narrowband communication, the UWB technique is robust against fading, and the multipath components can be resolved with a differential delay of less than a nano second. The resolved multipath can be well used in the Rake receiver.

In this chapter we focus on the UWB wireless channel. The results of this chapter are important to the designer of UWB communication system to predict the signal coverage, to estimate the maximum achievable data rate, to determine optimum location for antennas, to design efficient modulation schemes and to study associated signal processing algorithms. In UWB communication if a narrow time transmitted pulse $w_{tr}(t)$ is sent through the channel, the received UWB signal $r(t)$ can be written as:

$$r(t) = Aw_{\text{rec}}(t - \tau) + n(t) \tag{4.1}$$

Where A is the attenuation, τ is the propagation delay between transmit and receive and $n(t)$ is the additive noise. It should be noted that, due to the UWB

H. Nikookar, R. Prasad, *Introduction to Ultra Wideband for Wireless Communications*, Signals and Communication Technology,
DOI 10.1007/978-1-4020-6633-7_4, © Springer Science+Business Media B.V. 2009

channel, the transmit waveform changes from w_{tr} to w_{rec}, is delayed by τ, and is attenuated by A. Moreover, the received signal is corrupted by the noise.

It is worth mentioning that in the UWB channel characterization, the channel has the effects of the transceiver and antennas. These effects can be removed by using a reference LOS measurement, where the propagation channel can be considered to approximate free space loss, or by using the reference measurement at a reference distance (e.g., 1 m) between transmit and receive units.

UWB channel models can be either deterministic or statistical. Further, it can be carried out in both time and frequency domains. As will explained later in this chapter the main statistics for modeling in time domain are cluster distribution, cluster arrival distribution, amplitude fading, path loss and shadowing. While in frequency domain analysis the geometry and position of poles are the important parameters to be measured. Figure 4.1 gives a snapshot of the available methods and the respective parameters needed to characterize the UWB channel.

The structure of this chapter is as follows. In Section 4.1 we put special emphasize on the time domain impulse response modeling of the UWB channel, since the multipath medium can fully be described by its time impulse responses. In this section the distribution of fading (sub-section 4.1.1) and time of arrivals (sub-section 4.1.2) will be explained in more details. Other important channel parameters such as path loss, power delay profiles and rms delay spread are

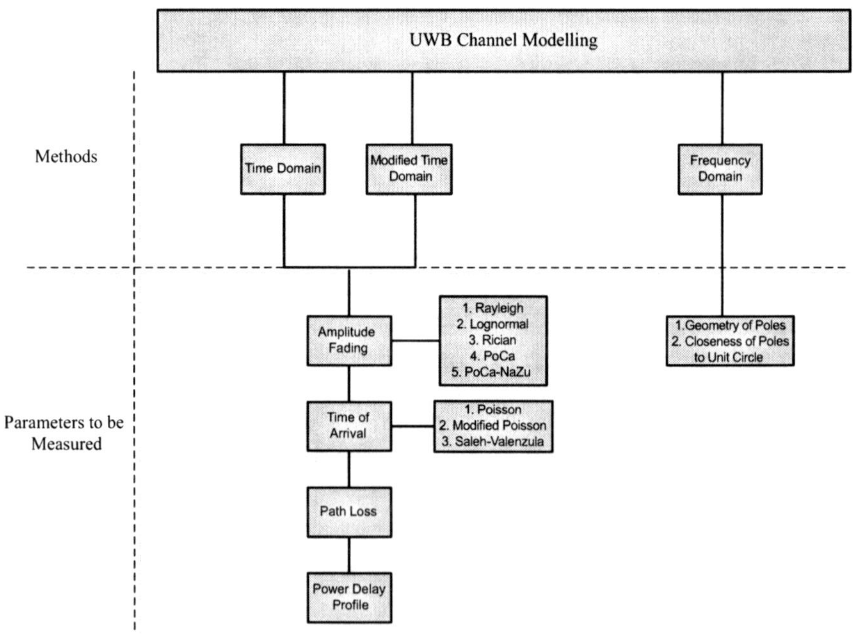

Fig. 4.1 UWB channel modeling – Methods and parameters

discussed in sub-sections 4.1.3–4.1.5. The modified impulse response method is discussed in Section 4.2. The UWB channel model proposed by IEEE 802.15.3 will be described in Section 4.3. In Section 4.4 the frequency domain modeling of the UWB wireless channels will be studied. The time and frequency modeling of UWB channels will be compared in Section 4.5. The chapter will be rounded up with a summary in Section 4.6.

4.1 Impulse Response Modeling of UWB Wireless Channels

The UWB wireless channel can be fully described by its impulse response function $h(t)$ which can be expressed as follows:

$$h(t) = \sum_{n=1}^{N} a_n \delta(t - \tau_n) e^{j\theta_n} \tag{4.2}$$

where the parameters of the *nth* path a_n , τ_n , θ_n and N are amplitude, delay, phase and number of multipath components, respectively. When UWB is a baseband signal, the phase in (4.2) can be kept out of consideration. The UWB wireless channel is completely characterized by these path variables. The mathematical model of (4.2) can be used to obtain the response $y(t)$ of the channel to any transmitted signal $s(t)$ by the convolution integral and adding noise,

$$y(t) = \int_{-\infty}^{\infty} s(x) h(t - x) dx + n(t) \tag{4.3}$$

where, $n(t)$ is the low pass complex value additive Gaussian noise.

Deduction to the Narrowband Model

By transmission of a constant envelope signal (i.e., $s(t) = 1.\exp(j\omega_0 t)$), through the multipath wireless channel, and ignoring the noise, the lowpass version of the channel output is

$$A\ e^{j\phi} = \sum_{k=0}^{N-1} a_k\ e^{j\theta_k} \tag{4.4}$$

This vector addition of the multipath components is represented in Fig. 4.2. According to (4.5) the received signal $y(t)$ is written as

$$y(t) = \mathrm{Re}[A\ e^{j\phi} e^{j\omega_0 t}] = A \cos(\omega_0 t + \phi) \tag{4.5}$$

In this case the effect of the multipath fading of the channel on the transmitted signal is changing the amplitude by a factor

Fig. 4.2 Vector summation of the multipath components in the narrowband model

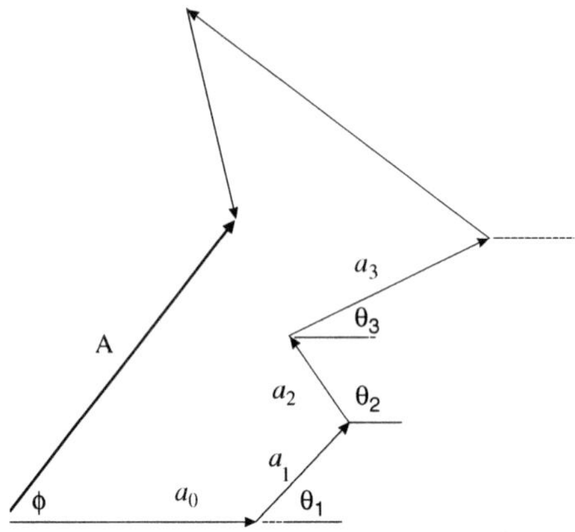

$$A = \sqrt{\left(\sum_{i=0}^{N-1} a_i \cos\theta_i\right)^2 + \left(\sum_{i=0}^{N-1} a_i \sin\theta_i\right)^2} \tag{4.6}$$

and phase by

$$\phi = \tan^{-1}\left(\frac{\sum_{i=0}^{N-1} a_i \sin\theta_i}{\sum_{i=0}^{N-1} a_i \cos\theta_i}\right) \tag{4.7}$$

4.1.1 Distribution of Amplitude Fading

In this sub section we focus our attention on the amplitude fading of UWB wireless propagation channels and provide a statistical model for signal amplitude fading. The amplitude fading in UWB wireless multipath environments may follow different distributions depending on area covered by measurements, presence or absence of a dominating strong component, and some other conditions. Major candidate distributions are as follows:

a) The Rayleigh Distribution

For narrowband communications, a common assumption that can be considered is that a_ks (scatter vectors in (4.4)) are nearly equal in length but have random phases. This assumption is well justified in small areas and in the absence of line of sight (LOS) component. Therefore, the vector summation of (4.4) is written:

$$r\,e^{j\theta} = r'\sum_{i=0}^{N-1} e^{j\theta_i} \tag{4.8}$$

where, $r' = a_0 \cong a_1 \cong \ldots \cong a_{N-1}$. The phase θ_i is very sensitive to the path length and changes by 2π as the path length changes by a wavelength. Therefore, phases have uniform distribution between $[0,2\pi)$. Considering this distribution for phases in (4.4), the joint distribution of r and θ will be the product of uniform $[0,2\pi)$ distribution with Rayleigh distribution described by

$$f(r) = \frac{r}{\sigma^2}\,\mathrm{e}^{-\frac{r^2}{2\sigma^2}}\,u(r) \tag{4.9}$$

where, $\sigma^2 = \Sigma\, a_i^2$ $(i = 0,1,2,\ldots,N-1)$, and $u(.)$ is the unit step function. The mean and variance of Rayleigh distribution are $\sqrt{(\pi/2)}\sigma$ and $(2-\pi/2)\sigma^2$, respectively. The assumption of approximately equal a_is in (4.4) may be unrealistic in practice. If these components are not equal but each individual component does not have a main contribution in received power (i.e., $a_i^2 << \Sigma_i\, a_i^2$), and when the number of components N is large, the envelope r in (4.4) will have Rayleigh distribution described by (4.9).

b) The Lognormal Distribution

Suppose x is a normal (Gaussian) random variable with mean μ and variance σ^2. If $x = \ln r$, it can be shown [1] that r has lognormal distribution as

$$f(r) = \frac{1}{\sigma r\sqrt{2\pi}}\,\mathrm{e}^{\left(-\frac{(\ln r-\mu)^2}{2\sigma^2}\right)}\,u(r) \tag{4.10}$$

where, $u(.)$ is the unit step function. A justification for using this distribution is when the fading is modeled as a multiplicative process. In this case, logarithms are added and according to Central Limit Theorem, the distribution will be lognormal.

c) The Rician Distribution

Referring to the Rayleigh distribution if we assume one of the components (r_k, θ_k) is fixed and other scatter vectors are random in amplitude and phase, (4.4) is rewritten as

$$r\,\mathrm{e}^{j\theta} = \alpha\,\mathrm{e}^{j\theta_0} + \sum_{i=0,\, i\neq k}^{N-1} a_i\,\mathrm{e}^{j\theta_i} \tag{4.11}$$

and probability density function of r will be [2]

$$f(r) = \frac{r}{\sigma^2}\,\mathrm{e}^{\left(-\frac{r^2+\alpha^2}{2\sigma^2}\right)}\,I_0\!\left(\frac{\alpha r}{\sigma^2}\right)u(r) \tag{4.12}$$

which is called the Rician distribution. In (4.11), α is a fixed vector, $u(.)$ is the unit step function, σ^2 is power of scatter Rayleigh components, and $I_0(\cdot)$ is the modified Bessel function of zeroth order

$$I_0(x) = \frac{1}{2\pi}\int_{-\pi}^{\pi} \mathrm{e}^{x\cos\phi}\,\mathrm{d}\phi \tag{4.13}$$

The Rician distribution occurs when a strong path exists in addition to the low level scattered path. The strong or fixed path component is line of sight path or the path that has much less attenuation with respect to other components. As a special case when $\alpha = 0$ (or $\alpha^2/2\sigma^2 \langle\langle r^2/2\sigma^2$) Rice distribution in (4.12) becomes Rayleigh distribution. Meanwhile, when the fixed vector has much more power than scatter components, r in (4.11) is approximately Gaussian. That is, in this case, the Rician distribution is approximated by a Gaussian distribution [3].

d) The PoCa Distribution

For the UWB communications because of very large bandwidth the number of sub-paths components in each bin (which is inversely proportional with the bandwidth), becomes very small. Accordingly, the Rayleigh and lognormal distributions do not seem to describe properly the distribution of fading of the UWB channels. Poydoro and Capsalis (PoCa) [4] proposed a distribution, known as PoCa distribution, for the fading of UWB channels. This distribution accurately describes the amplitudes when the number of sub-paths in each bin is small. If n denotes the number of sub-paths in each bin and σ^2 the variance of the Rayleigh distribution, the PoCa distribution can be expressed as:

$$\begin{aligned} f(r) &= \frac{2\pi K^2(n)}{\sigma^2}\frac{r}{(a^2-b^2)^{\frac{n+1}{2}}}P_n\left(\frac{a}{\sqrt{a^2-b^2}}\right) \\ a &= 1+\frac{r^2}{\sigma^2(2n+1)}+\frac{r^4}{8\sigma^4(2n+1)^2} \\ b &= \frac{r^4}{8\sigma^4(2n+1)^2} \\ K(n) &= \frac{1}{\sqrt{\pi(2n+1)}}\frac{\Gamma(n+1)}{\Gamma(n+\frac{1}{2})} \end{aligned} \tag{4.14}$$

where

$$\Gamma(n) = \int_0^\infty u^{n-1}\,\mathrm{e}^{-u}\mathrm{d}u \tag{4.15}$$

and $P_n(x)$ denotes the Legandre function*. In Fig. 4.3 the probability distribution function (PDF) of the PoCa distribution for different number of subpaths components as well as the Rayleigh distribution is depicted. From this figure a noticeable difference between the Rayleigh and PoCa distribution (with small n)

---

* Legandre polynomials for several n are as follows: $P_0(x)=1$, $P_1(x)=x$, $P_2(x)=-(3x^2-1)$, $P_3(x)=(5x^3-3x)/2, \ldots$

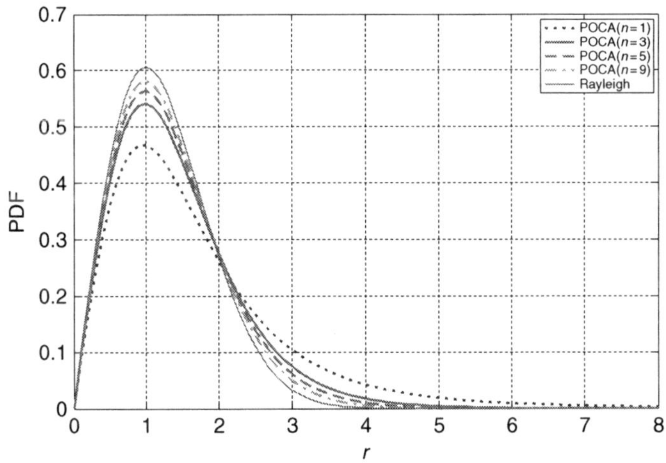

Fig. 4.3 The PDF of PoCa distribution for different number of subpaths

is seen. When the number of subpaths gets larger the PoCa distribution approaches the Rayleigh distribution.

e) PoCa-NaZu Distribution

We mentioned earlier that the Ricean distribution is a good distribution to specify the amplitude fading when LOS is present. For the UWB channels with a strong LOS and small number of subpaths in each bin, the PoCa-NAZU (Nakagawa-Arita-Zhang-Udagawa) distribution is suggested to describe the amplitude fading [5]. This distribution is given by:

$$
\begin{aligned}
x_0 &= 1 + \frac{s^2 + r^2}{(2n+1)\sigma^2} + \frac{s^2 r^2}{(2n+1)^2 \sigma^4} \\
x_1 &= \frac{-2sr}{(2n+1)\sigma^2} - \frac{2sr^3}{(2n+1)^2 \sigma^4} \\
x_2 &= \frac{-s^2 r^2 + r^4}{(2n+1)^2 \sigma^4} \\
x_3 &= \frac{2sr^3}{(2n+1)^2 \sigma^4} \\
x_4 &= \frac{-r^4}{(2n+1)^2 \sigma^4}
\end{aligned}
\tag{4.16}
$$

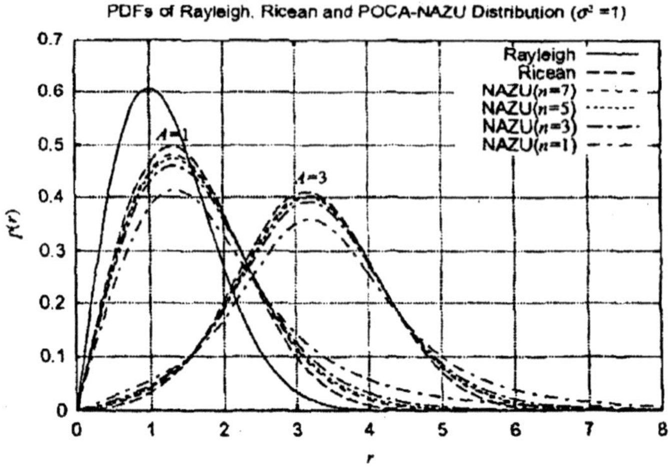

Fig. 4.4 The PDF of Rayleigh, Ricean and PoCa-NAZU distribution for different number of subpaths, courtesy of [5]

where

$$f(r) = r\frac{K^2(2n+1)}{\sigma^2}\int_0^{2\pi}(x_0 + x_1\cos u + x_2\cos 2u + x_3\cos 3u + x_4\cos 4u)^{-(n+1/2)}\mathrm{d}u \tag{4.17}$$

In Fig. 4.4 the Rayleigh, Ricean distributions as well as the PoCa-NAZU (for different values of n) are illustrated. As seen from the figure for large n the PoCa-NAZU distribution tends to the Ricean distribution.

4.1.2 Distribution of Time of Arrival

a) Poisson Distribution

Distribution of the arrival time sequence (i.e., τ_n in Eq. (4.2)) has been investigated in [6]–[9]. As a first model Poisson distribution for the sequence of path arrival times may be considered. This distribution is often addressed when certain events occur with complete randomness. If I denotes the number of paths occurring in a given time interval T, Poisson distribution will be

$$P(I=i) = \frac{\mu^i\ \mathrm{e}^{-\mu}}{i!} \tag{4.18}$$

where $\mu = \int_T \lambda(t)\mathrm{d}t$ and $\lambda(t)$ is mean arrival rate at time t. For a stationary process $\lambda(t)$ is constant and $\mathrm{E}[I] = \mathrm{Var}(I) = \lambda$. Analysis of time of arrival of multipath components of the indoor and mobile data base has shown that standard Poisson model does not provide a good fit.

b) Modified Poisson Distribution

A more realistic model is the "modified Poisson" or Δ-K model. This model which takes into account the clustering properties of multipath components was first suggested by G.L. Turin [6] and was successfully used in analysis [7] and simulation [8]–[3] of mobile and indoor radio propagation channels. In standard Poisson model path components are in complete randomness, however, in the modified Poisson process, occurring of a path will change the probability of having another one. As shown in Fig. 4.5 the process starts with a pure Poisson. If a path exists at time t then the process will switch to another Poisson process with parameter $K\lambda(t)$ and if there is no further path in the interval $[t,t+\Delta)$ it comes back to its initial state at the end of the interval. This model is described by a series of transitions between two states. With $K=1$ and $\Delta=0$ this process reverts to the standard Poisson process. The application of discrete version of this process to the path arrival times of measured data of mobile and indoor radio propagation channels shows a very good fit. (References [7]–[8] and [9], [3], respectively). The good fit of Δ-K model in describing the time of arrival of path components is mainly due to nonrandomness of local structure and objects, which means multipath components occurr in groups. Another justification is that the modified Poisson due to its nature, uses more information of the data compared to pure Poisson, i.e., the Δ-K model uses empirical probabilities associated with individual small intervals Δ, while standard Poisson process model uses the total probability associated with a larger interval $T >> \Delta$, [7], [8].

c) Saleh and Valenzula Model – Double Poisson Distribution

Another method to characterize the arrival times in UWB channels is the double Poisson model. This model was first proposed by Saleh and Valenzula (SV) for indoor channels [10]. According to this model multipath arrivals occur

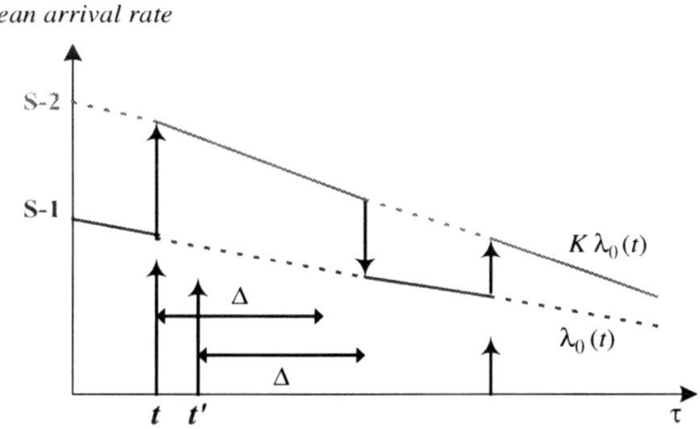

Fig. 4.5 The continuous time Modified Poisson Process

in clusters. The rate of cluster arrivals is Λ. Within each cluster, rays (multipath) arrive according to Poisson Process with rate λ. When the arrival process is Poisson the inter-arrival times are exponentially distributed. If T_l denotes the arrival time of the lth cluster we have:

$$\mathrm{Prob}(T_1|T_{l-1}) = \Lambda\ \mathrm{e}^{-\Lambda(T_1 - T_{l-1})} \tag{4.19}$$

and for the nth arrival in the lth cluster we will have:

$$\mathrm{Prob}(\tau_{n,l}|\tau_{n,l}) = \lambda\ \mathrm{e}^{-\lambda(\tau_{n,l} - \tau_{n-1,l})} \tag{4.20}$$

Accordingly, the impulse response of the UWB channel becomes:

$$h(t) = \sum_{l=0}^{\infty}\sum_{n=0}^{\infty} a_{n,l}\delta(t - T_l - \tau_{n,l})\ \mathrm{e}^{j\theta_{n,l}} \tag{4.21}$$

where $a_{n,l}$, $\theta_{n,l}$ are respectively the amplitude and phase of the nth ray in the lth cluster. With this model the power delay profile can be expressed by two negative exponential functions as

$$P_{n,l} = P_{0,0}\ \mathrm{e}^{-T_1/\Gamma}\ \mathrm{e}^{-\tau_{n,l}/\gamma} \tag{4.22}$$

where $P_{0,0}$ is the received power at delay 0 of the 0th cluster. The parameter γ and Γ are the ray and cluster time decay constants (TDC) of the power delay profiles, respectively. Figure 4.6 shows several power decay profiles with exponential decaying clusters and rays. In Table 4.1 the parameters of double exponential power decay profiles for the UWB as well as wideband systems are depicted, [11]–[13].

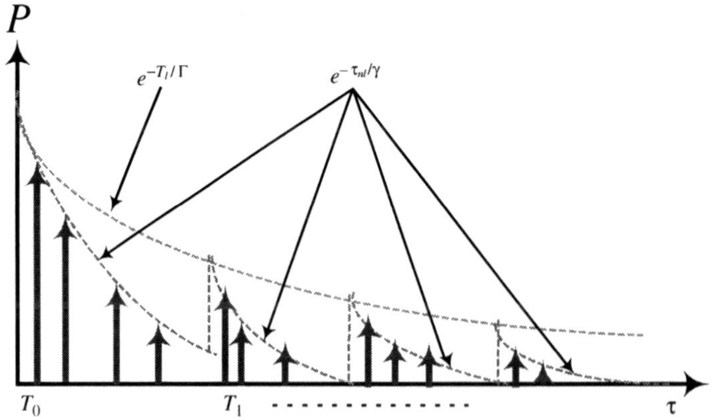

Fig. 4.6 Several power delay profiles with exponential decaying cluster and rays

Table 4.1 Double exponential model for the UWB and conventional wideband systems

Parameter	UWB		Conventional wideband		
	Win Model [11]	Intel Model [12]	Saleh-Valenzuela[10]	Spencer et al. [13]	
				Building 1	Building 2
TDC of clusters (ns)	27.9	16	60	33.6	78
TDC within cluster (ns)	84.1	1.6	20	28.6	82.2
Cluster arrival rate (1/ns)	1/45.5	1/60	1/300	16.8	17.3
Intracluster arrival rate (1/ns)	1/2.3	1/0.5	1/5	5.10	6.6

4.1.3 Path Loss

Path loss or path attenuation, by definition is the attenuation undergone by an electromagnetic wave in the transit between a transmitter and a receiver in a communication system. Path loss may be due to many effects such as free space path loss, reflection, refraction, diffraction, scattering, clutter, absorption from objects, structures or any other obstructions in the path.

Path loss (*PL*) is defined as:

$$PL = \frac{P_t}{P_r} \tag{4.23}$$

And in log scale:

$$PL(dB) = P_t(dB) - P_r(dB) \tag{4.24}$$

where P_t is the power fed to the transmitting antenna and P_r is the power available at the receiving antenna. For very simple cases where there is a direct path from transmitter to receiver, and in the absence of substantial obstacles in the path of the signal, the received power follows the inverse square law:

$$P_r \propto d^{-2}, \tag{4.25}$$

where *d is* the distance between transmitter and receiver. Narrowband signals propagate in free space as three-dimensional expanding waves. Beyond the immediate near-field vicinity of the antenna, propagating energy expands spherically in proportion to the square of the distance. For omni-directional expansion, the total energy remains constant over the surface area $4\pi d^2$ of a sphere having radius d. According to Friis law free space loss is:

$$PL = 20\log_{10}\left(\frac{4\pi d f_c}{c}\right) \tag{4.26}$$

where d is the distance, c is the speed of light and f_c is the center frequency. The received power at distance d, $P_r(d)$ is expressed as:

$$P_r = P_t G_t G_r \left(\frac{c}{4\pi f d}\right)^2, d>0 \tag{4.27}$$

Where P_t is the transmit power, and G_t and G_r are the transmit and receive antenna gains, respectively. Some losses have to be taken into account in (4.27) such as filter and antenna losses that are not considered here. In logarithm scale (4.27) is written as:

$$P_r = P_t + G_t + G_r + 20\log\left(\frac{c}{4\pi f d}\right) \tag{4.28}$$

Using (4.23) the path loss in free space is given as:

$$PL(d) = \frac{(4\pi)^2 d^2 f_c^2}{G_r G_t c^2} \tag{4.29}$$

A common measure of the average large-scale path loss for a given separation d between the transmitter and receiver is expressed by:

$$PL(d) \propto \left(\frac{d}{d_0}\right)^n \tag{4.30}$$

where n is path loss exponent and d_0 is a reference distance. The reference distance must be smaller than the typical distances encountered in wireless communication systems and must fall in the far field region of the antenna, so that the losses beyond that point are purely distance dependent effects. In log scale (4.30) can be written as:

$$PL(d) = PL(d_0) + 10n\log\left(\frac{d}{d_0}\right) \tag{4.31}$$

For the UWB channels the path loss formula (4.29) may be used by replacing center frequency f_c with the geometrical mean of the f_l and f_h (the low and high frequency band edges of the UWB signal, respectively). But care should be taken as the wavelength (which is c/f) will be different for different frequencies of the UWB band. Meanwhile, because of very large bandwidth of UWB signal, the gain of the antennas varies with the frequency. For the UWB propagation channel, different path loss exponents are reported in the literature. The path

Table 4.2 Breakpoint distance and path loss exponents for different bandwidths

Breakpoint Distance and Path loss Exponent			
Bandwidth	Breakpoint Distance	Path Loss Exponent	
		$d < D_b$	$d > D_b$
7.5 GHz	97 m	1.8	4.0
750 MHz	72 m	1.9	4.0
75 MHz	65 m	2.0	4.0
7.5 MHz	64 m	2.0	4.0
CW	64 m	2.0	4.0
Lowest Frequency = 3.1 GHz			

loss exponent depends on the environment. In [14] for LOS UWB indoor channels a path loss of 1.7 and for NLOS case a path loss of 3.5 is reported. A dual slope path loss model for narrowband as well as UWB channels is proposed in [15]. According to this model for the distances lower than the beak point distance D_b the path loss exponent is 1.8 and above this distance is 4. The breakpoint distance depends on the heights of transmitting and receiving antennas, the lowest frequency as well as the bandwidth of the signal (see Table 4.2).

Another important issue is the dependence of path loss with the frequency. Generally path loss increases with the frequency. In [16] for a 2-path channel it is shown that the path loss formula (4.31) can be expressed as a function of distance and frequency:

$$PL(d,f) = PL(d_0,f_0) + 10n\log\left(\frac{d}{d_0}\right) + 10m\log\left(\frac{f}{f_0}\right) \tag{4.32}$$

where d_0 and f_0 are the reference distance and frequency, respectively. Referring to (4.29) for the free space path loss $m = 2$. For the UWB indoor channels a dependence of path loss exponent with the carrier frequency and bandwidth is reported in [17]. For a corridor scenario in LOS the path loss exponent ranges 1.55–2.2 and as a general trend it increases with the increasing carrier frequency. For NLOS, path loss exponent ranges 2–6.5. According to [18] for NLOS measurements, path loss exponent increased from 3.3 to 4.5 when frequency increased from 2.4 to 11.5 GHz.

4.1.4 Power-Delay Profiles

Using the impulse response of the channel (i.e. Eq. (4.2)), the received signal power can be obtained as

$$P(\tau_n) = E\{|h(t)|^2\} = \sum_{n=0}^{N-1} \overline{a_n{}^2}\,\delta(t-\tau_n) \tag{4.33}$$

Usually the later paths of the power delay profile experience more attenuation and accordingly the power delay profiles are generally decreasing function of the excess delay. A good model is exponential decrease of the received power as:

$$P(\tau_n) = P(0)\sum_{n=0}^{N-1} e^{-\alpha\tau_n}\delta(t-\tau_n) \tag{4.34}$$

where $P(0)$ is the received power at delay τ_0 and α is a parameter that controls the decreasing shape of the power delay profile. Higher values of α give faster decaying of the power delay profiles. The indoor measurements results of [19] and [20] show the exponential decrease of the power delay profiles with $1/\alpha = 29$ ns for the LOS and $1/\alpha = 40$ ns for the NLOS case.

4.1.5 RMS Delay Spread

For the UWB wireless channels a one-number representation of an impulse response profile is the rms delay spread. The rms delay spread (τ_{rms}) is defined as:

$$\tau_{rms} = \sqrt{\overline{\tau^2} - (\overline{\tau})^2} \tag{4.35}$$

where the mean excess delay $\overline{\tau}$ is the first moment of the power delay profile and is given by:

$$\overline{\tau} = \frac{\sum_i a_i^2\tau_i}{\sum_i a_i^2} = \frac{\sum_i P(\tau_i)\tau_i}{\sum_i P(\tau_i)} \tag{4.36}$$

The second moment of the power profile and is defined as:

$$\overline{\tau^2} = \frac{\sum_i P(\tau_i)\tau_i^2}{\sum_i P(\tau_i)} \tag{4.37}$$

Rms delay spread is a good measure of multipath spread of the channel. Strong echoes with long delays contribute significantly to τ_{rms}. It has been shown that the performance of UWB communication systems working in the wireless channels is very sensitive to the value of τ_{rms}. The results of UWB measurements show a mean of 4.7 ns for the rms delay spread of the LOS indoor channels and a mean of 8.2 ns for NLOS [14]. In [20] τ_{rms} values of 8–12 and 14–19 ns are reported for LOS and NLOS cases, respectively.

4.2 Modified Impulse Response Method

After discussion of impulse response method of UWB channels we are in a position to focus on the Eq. (4.2), as in the UWB case due to very large bandwidth the channel model expressed by this equation may need some modifications to proper characterize the channel. The Fourier transform of $h(t)$ in (4.2) is written as:

$$FT\{h(t)\} = H(f) = \sum_{n=1}^{N} a_n \mathrm{e}^{j(2\pi f \tau_n + \theta_n)} \tag{4.38}$$

According to the impulse response model of (4.2), each path is independent of frequency. This assumption is valid only when the bandwidth around the center frequency is not large (the condition that is not met for the UWB case). Meanwhile, for the ultra large bandwidths the scattering and diffraction, that cause multipath components in the impulse response model, are strongly frequency dependent. Therefore, for a precise impulse response modeling of the UWB channels, the frequency dependence of the individual paths should be incorporated in the model. This can be done by adding a frequency dependent term to the Eq. (4.38) as:

$$H_{UWB}(f) = \sum_{n=1}^{N} a_n g_n(f) \mathrm{e}^{j(2\pi f \tau_n + \theta_n)} \tag{4.39}$$

The frequency dependent term $g_n(f)$ can have different values. For example for the direct path that can be 1, for the diffraction from wall edge can be represented by $f^{-0.5}$, for the diffraction from desk corner can be expressed by f^{-1}, or for the case of double diffraction can be considered as $f^{-1.5}$, [21]. The other example of frequency dependence is when more than one (e.g., 3) ray arrive in the same time. The contribution of three rays is equivalent to one ray with frequency dependent amplitude:

$$g_n(f) = 1 + \frac{a_{n'}}{a_n} \mathrm{e}^{j2\pi f(t_n - t_{n'})} + \frac{a_{n''}}{a_n} \mathrm{e}^{j2\pi f(t_n - t_{n''})} \tag{4.40}$$

For slowly varying $g_n(f)$ we can expand that around center frequency f_0 as:

$$g_n(f) = g_n(f_0) + g_n'(f_0)(f - f_0) + \frac{1}{2} g_n''(f_0)(f - f_0)^2 + \dots \tag{4.41}$$

Accordingly, the frequency response of the UWB channel can be written as:

$$H_{\mathrm{UWB}}(f) = \sum_{n=1}^{N} a_n g_n(f) \mathrm{e}^{j2\pi f \tau_n} \cong \sum_{n=1}^{N} A_n \mathrm{e}^{j2\pi f T_n}$$
$$T_n = \tau_n + j\frac{g'_n(f_0)}{g_n(f_0)} \tag{4.42}$$

T_n is considered as the "complex path arrival time". Contemplating (4.42) we notice that if the bandwidth is so large that the frequency dependence of multipath should be considered then the model of path delays will not be simply τ_n. But rather the imaginary part (which is a function of frequency dependent term added to the channel frequency response) should be considered as well. The larger the bandwidth the stronger is the effect of frequency dependency. Based on the modified impulse response method, another point that should be noted here is the rms delay spread of the UWB channels. Using definition of τ_{rms} in (4.35)–(4.37), it is seen from (4.42) that for the UWB channels the rms delay spread is a function of the frequency.

4.3 The IEEE UWB Channel Model

IEEE 802.15.3a proposed a UWB multipath channel model in July 2003, [22]. It is based on the S-V model explained in the previous section. The multipath components arrive at the receiver in groups, called clusters, with Poisson distribution. The path (ray) within each cluster also arrives with Poisson distribution. The channel impulse response is given by:

$$h(t) = X \sum_{l=1}^{L} \sum_{n=1}^{M} \alpha_{nl} \delta(t - T_l - \tau_{nl}) \tag{4.43}$$

where L is number of clusters, M is number of paths within a cluster, $\alpha_{n,l}$ is the multipath gain of the nth path corresponding to lth cluster. T_l is delay of lth cluster and $\tau_{n,l}$ is the time delay of nth ray of the lth cluster. The amplitude fading is defined as $\alpha_{n,l} = P_{n,l} \cdot \zeta_l \, \beta_{n,l}$, where $\mathrm{P}_{n,l}$ is the sign of the coefficient and takes ± 1 with equal probability and accounts for signal inversions due to reflections. ζ_l is the fading associated with the lth cluster. $\beta_{n,l}$, is the fading associated with the nth ray of the lth cluster. The envelope of amplitude fading $\alpha_{n,l}$ (i.e., $\zeta_l \beta_{n,l}$) is lognormally distributed with mean $\mu_{n,l}$ and variance $\sigma_1{}^2 + \sigma_2{}^2$. Using this model:

$$E\{|\zeta_l \beta_{n,l}|^2\} = \Omega_0 e^{-T_l/\Gamma} e^{-\tau_{n,l}/\gamma} \tag{4.44}$$

where, Ω_0 *is* the mean power of the first path of the first cluster at delay 0 of the first cluster. Parameters Γ and γ are the cluster and ray decay constants, respectively. According to this model, the mean of the channel fading is given by:

$$\mu_{n,l} = \frac{10 \ln \Omega_0 - 10 T_l/\Gamma - 10\tau_{n,l}/\gamma}{\ln 10} - \frac{\ln 10}{20}\left(\sigma_1{}^2 + \sigma_2{}^2\right) \tag{4.45}$$

where $\sigma_1 = \sigma_2 = 3.3941$ dB. Since the clusters and rays (within each cluster) arrive according to Poisson process, the inter arrival times of the clusters and rays follow the exponential distributions as:

$$\begin{aligned} \text{Prob}(T_l | T_{l-1}) &= \Lambda e^{-\Lambda(T_l - T_{l-1})} \\ \text{Prob}(\tau_{n,l} | \tau_{n-1,l}) &= \lambda e^{-\lambda(\tau_{n,l} - \tau_{n-1,l})} \end{aligned} \tag{4.46}$$

Where Λ is the cluster arrival rate and λ is the ray arrival rate.

Parameter X in (4.43) is the shadowing which can be modeled by lognormal distribution. The IEEE 802.15.3.a considers 4 channel models (CM1-CM4). In the CM1 the transmit-receive antenna separation is 0–4 m and the situation is LOS. The CM2 is NLOS, and the transmit-receive separation is within 4 m. The channel model CM3 is for NLOS, with the transmit-receive separation of 4–10 m. The channel model CM4 is for NLOS extreme multipath with rms delay spread above 25 ns, [22]. In Table 4.3 the IEEE UWB channel model parameters for CM1 and CM3 are shown.

Figure 4.7 gives the methods and respective parameters used to characterize the IEEE UWB channel model.

Table 4.3 IEEE UWB Channel model parameters

Model Parameter	CM1	CM3	Unit
LOS/NLOS	LOS	NLOS	–
TX-RX Separation	0–4	4–10	m
Λ	0.0233	0.0667	1/ns
λ	2.5	2.1	1/ns
Γ	7.1	14	ns
γ	4.3	7.9	ns
σ_1	3.3941	3.3941	dB
σ_2	3.3941	3.3941	dB

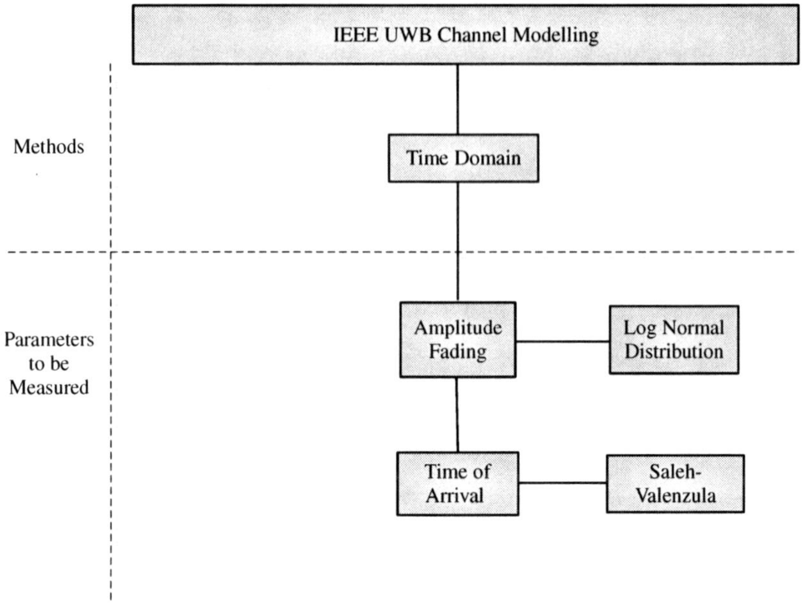

Fig. 4.7 IEEE UWB channel modeling – Methods and parameters

4.4 Frequency Modeling of UWB Channels

Another approach for the modeling of mobile/indoor radio propagation channel is using the frequency domain. The main advantage of the modeling in the frequency domain is that it requires lower number of parameters for characterization of the channel than time domain methods. Full description of the frequency domain modeling of wireless channels are reported in [23] and [24]. The basic idea of this technique is that the frequency response of the channel at each physical point, i.e., $H(f)$, is modeled by an auto-regressive (AR) process of order p as

$$H(f_n) = \sum_{i=1}^{p} a_i H(f_{n-i}) + V(f_n) \tag{4.47}$$

where $H(f_i)$ is the nth sample of the complex frequency domain measurement and $V(f_n)$ is a complex white noise process representing the error between the actual frequency response value at frequency f_n and its estimate based on the last p samples of the frequency response [24].

$$G(z) = \frac{1}{1 - \sum_{i=1}^{p} a_i z^{-i}} = \prod_{i=1}^{p} \frac{1}{(1 - p_i z^{-1})} \tag{4.48}$$

Using this model the frequency response of the wireless channel is determined by the output of a linear filter with the transfer function $G(z)$, having p poles when excited by a white noise process with the variance

$$\sigma^2 = R(0) - \sum_{i=1}^{p} a_i R(i) \tag{4.49}$$

where a_is are the parameters of the model and are solutions of the Yule-Walker Equations:

$$R(-n) = \sum_{i=1}^{p} a_i R(i-n), \qquad 1 < n < p \tag{4.50}$$

and $R(\cdot)$ is the auto correlation function of the frequency response as

$$R(k) = \frac{1}{L} \sum_{i=1}^{L-k} H^*(f_i) H(f_{i-k}), \qquad k \geq 0 \tag{4.51}$$

where L is the number of points used in the frequency domain measurement. In the AR modeling the frequency response of the UWB wireless channel represented by L measurement samples is described by p parameters of the model where typically $p << L$. Reported results of [14] indicate that the UWB indoor channel can be well modeled in the frequency domain by a linear filter with the transfer function having only 2 significant poles (or 2 significant clusters). The block diagram of the AR filter (with two poles) for modeling of UWB channel is shown in Fig. 4.8. The geometry of the poles is important. The delay associated with a pole is determined by the angle of that pole and the sampling frequency f_s as

$$\tau_i = \text{angle}\{p_i\} \,/\, 2\pi f_s \tag{4.52}$$

The closeness of a pole to the unit circle represents significant power at the corresponding delay.

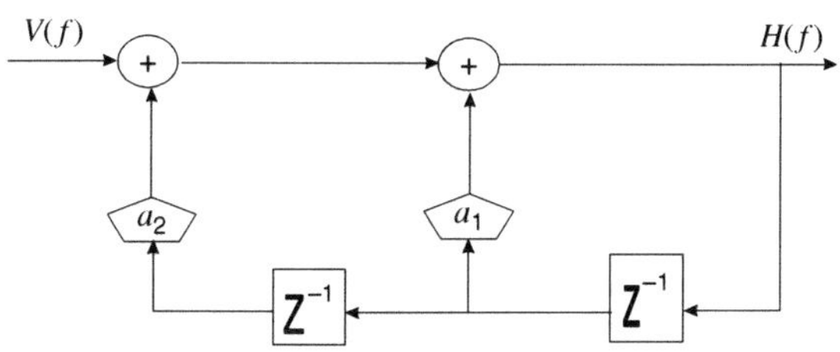

Fig. 4.8 Block diagram of the AR modeling of the UWB channel

4.5 Comparison of Time and Frequency Models

The impulse response time domain approach is by far the most popular approach used for modeling wideband indoor and urban radio channels as well as UWB channels. In the impulse response modeling of wireless channels there is a straightforward relationship between multipath propagation and the results of measurements. With the frequency modeling the physical interpretation of the model is not so straightforward, because the path arrivals are not directly related to the results of frequency-domain measurements.

Computation of parameters of the AR model is straightforward and generally accurate. While in time domain they are rather tedious and prone to inaccuracy [23]. The implementation of time domain modeling of wireless channels is more complicated than the frequency domain AR model. In the simulation of UWB wireless channels based on the impulse response model, the double Poisson (or modified Poisson) process as well as random variables with proper distributions for the amplitude fading and phases should be generated (see sub sections 4.1.1–4.1.2). This is more complicated than simulation of an AR model that uses a filter block and a fast Fourier block – to generate the frequency and time domain responses.

4.6 Summary

In order to ensure effective transmission of UWB signals it is important to understand and develop a model of the channel that adequately describes the UWB environment. Developing of a channel model for the UWB environments is very challenging, particularly so because of the very high bandwidth of UWB transmission [26]. The knowledge of the channel helps to a huge extent in the proper design of wireless communication system and evaluation of performance.

Further, such analysis reduces the cost of developing a complex system by reducing the amount of hardware that has to be developed for evaluation of performance. This chapter focused on UWB wireless channel. The important parameters including amplitude fading, time delay, RMS delay spread based on which the channel models are devised were explained. IEEE802.15.3 which is a popular UWB standard was presented. The results of this chapter are important to the designer of UWB communication system to predict the signal coverage, to estimate the maximum achievable data rate, to determine optimum location for antennas, to design efficient modulation schemes and to study associated signal processing algorithms.

Problems

Problem 4.1 Using the impulse response model for a two-path channel with $\{a_0 = 1,\ \tau_0 = 0,\ \theta_0 = 0\}$ and $\{a_1 = a,\ \tau_1 = T$ and $\theta_1 = \theta\}$ obtain the frequency response of the channel, and show that for $a = 1$ it can be written as

$$|H(f)| = 2\left|\cos\left(\pi f T - \frac{\theta}{2}\right)\right|$$

(a) Sketch the frequency response of the channel for $\theta = 0$.
(b) Show that the rms delay spread of the channel is $T/2$. Now assume that the path amplitudes a_0 and a_1 are Rayleigh distributed with power ratio of $\gamma = \sigma_0^2/\sigma_1^2$. Calculate the rms delay spread of the channel as a function of γ and plot it for $\gamma = 0.1$, $\gamma = 0.5$ and $\gamma = 1$.
(c) Repeat parts (a) and (b) for 3 equal power path components.
(d) What is the effect of path phase distribution on the rms delay spread?

Problem 4.2 In this problem we would like to study the impact of power-delay profile shape on the rms delay spread of the UWB wireless channels. Consider 3 types of UWB channels with the following power delay profiles:

$$P_A(\tau) = \begin{cases} P_0 e^{-\alpha\tau} & 0 \le \tau \le T_{\max} \\ 0 & \text{otherwise} \end{cases}$$

$$P_B(\tau) = \begin{cases} P_1 e^{-\beta\tau} & 0 \le \tau \le T_{\max} \\ 0 & \text{otherwise} \end{cases}$$

$$P_c(\tau) = \begin{cases} P_2 e^{-\alpha\tau} & 0 \le \tau \le \frac{T_{\max}}{2} \\ P_2 e^{-\alpha(\tau - T\max/2)} & \frac{T_{\max}}{2} \le \tau \le T_{\max} \\ 0 & \text{otherwise} \end{cases}$$

where $T_{\max}$ is the maximum measurable delay spread of the channels, $\beta = 2\alpha$, and $\alpha^{-1} = T_{\max}/3$.

(a) Calculate P_1 and P_2 in terms of α and $T_{\max}$ in order that total average power of three channels be the same.
(b) Obtain the τ_{rms} of the three channels and compare them to each other. Conclude the effect of power delay profile shape on the rms delay spread of the channel.

Problem 4.3 The existence of excess time delays in the channel impulse response causes that the two closely spaced in frequency input signals to the channel become correlated out of the wireless channel. Let $H(f)$ denote the frequency transform of the impulse response $h(t)$. The coherence bandwidth is related to the frequency correlation function

$$R_H(\Delta f) = \frac{E\{H^*(f)H(f+\Delta f)\}}{E\{|H(f)|^2\}}$$

the coherence bandwidth B_c of level k is the smallest number so that $|R_H(B_c)| < k$. For a two-path channel, where both path have equal powers, show that the coherence bandwidth for having frequency correlation of k, can be written as

$$B_k = \frac{\cos^{-1}(k)}{2\pi\tau_{\rm rms}}$$

Problem 4.4 A UWB wireless channel has been modeled in frequency by 2 poles as: $P_1 = 0.9860\ \angle -18.5^\circ$, and $P_2 = 0.8614\ \angle -8.483^\circ$. The sampling frequency has been 5 GHz.

(a) Sketch the power delay graph of the channel using the location of poles. (Hint: Use the inverse Fourier transform of a transfer function with poles as specified above)
(b) Using (a) calculate the rms delay spread of the channel.

Problem 4.5 In this chapter we studied the SV model of UWB channels. The SV model uses a Poisson process for the ray arrival times. Due to discrepancy in the fitting of measured results of indoor/outdoor UWB channels, a mixture of two Poisson distributions is proposed in [25] as:

$$\text{Prob}(\tau_{n,l}|\tau_{n-1,l}) = \beta\lambda_1 e^{-\lambda_1(\tau_{n,l}-\tau_{n-1,l})} + (1-\beta)\lambda_2 e^{-\lambda_2(\tau_{n,l}-\tau_{n-1,l})}$$

Where β is the mixture probability and λ_1 and λ_2 are the ray arrival rates. For a UWB channel having only one cluster and ray arrival rates of $\lambda_1 = 5$ and $\lambda_2 = 2\ \text{ns}^{-1}$ and a power delay profile of the form

$$P(\tau_n) = \sum_{n=1}^{15} e^{-\tau_n/20\,ns}\delta(t-\tau_n)$$

(a) For $\beta = 1/2$, obtain the rms delay spread of the channel.
(b) Repeat part (a) if $\beta = 1/4$, $\beta = 1$and 0. What do you conclude?
(c) Which value of β minimizes the rms delay spread?

Problem 4.6 Starting from (4.43) and using the Taylor expansion of $g_n(f)$ around the center frequency derive (4.46).

References

1. A.U.H. Sheikh, Wireless Communications, Theory and Techniques, Kluwer, Boston, 2004.
2. S.O. Rice, "Mathematical analysis of random noise," Bell syst. Tech. J., vol. 23, pp. 282–332, 1954, and vol. 24, pp. 46–156, 1954.
3. H. Hashemi, "The indoor radio propagation channel," Proceedings of the IEEE, vol. 81, no. 7, July 1993, pp. 943–968.
4. D.S. Polydorou and C.N. Capsalis, "A new theoretical model for the prediction of rapid fading variations in an indoor environment," IEEE Transactions on Vehicular Technology, vol. 46, no. 3, August 1997, pp. 748–754.

5. H. Zhang, T. Udagawa, T. Arita and M. Nakagawa, "A statistical model for the small-scale multipath fading characteristics of ultra wideband indoor channel" IEEE Conference on Ultra Wideband Systems and Technologies, Digest of Papers, pp. 81–86, 2002.
6. G.L. Turin, F.D. Clapp, T.L. Johnston, S.B. Fine and D. Lavry, "A statistical model of urban multipath propagation," IEEE Transactions on Vehicular Technology, vol. VT-21, no. 1, February 1972, pp. 1–9.
7. H. Suzuki, "A statistical model for urban radio propagation," IEEE Transactions on Communications, vol. COM-25, July 1977, pp. 673–680.
8. H. Hashemi, "Simulation of urban radio propagation channel," IEEE Transactions on Vehicular Technology, vol. VT-28, August 1979, pp. 213–224.
9. H. Hashemi, "Impulse response modeling of indoor radio propagation channels," IEEE J. Select. Area Commun., vol. 11, no. 7, September 1993, pp. 967–978.
10. A.A.M. Saleh and R.A. Valenzuela, "A statistical model for indoor multipath propagation", IEEE J. Select. Areas Commun., vol. 5, 1987, pp. 128–137.
11. R.J.-M. Cramer, R.A. Scholtz and M.Z. Win, "Evaluation of an ultra-wide-band propagation channel",IEEE Transactions on Antennas and Propagation, vol. 50, no. 5, May 2002, pp. 561–570.
12. J. Foerster and Q. Li, "UWB Channel modeling," Contribution from Intel- Report, June 2002.
13. Q. Spencer, et al., "A statistical model for angle of arrival in indoor multipath propagation", IEEE 47th Vehicular Technology Conference, vol. 3, May 1997, pp. 1415–1419.
14. S.S. Ghassemzadeh, R. Jana, V. Tarokh, C.W. Rice and W. Turin, "A statistical path loss model for in-home UWB channels," in Proc. IEEE Conference Ultra Wideband Systems and Technologies, pp. 59–64, 2002.
15. K. Siwiak and H. Bertoni and S.M. Yano, "Relation between multipath and wave propagation attenuation," Electron Lett., vol. 39, January. 2003, pp. 142–143.
16. M. Ghavami, L.B. Michael and R. Kohno, Ultra Wideband Signals and Systems in Communication Engineering, Wiley, 2004
17. D. Cassioli, A. Durantini and W. Ciccognani, "The role of path loss on the selection of the operating bands of UWB systems," in Proc. IEEE Int. Symp. on Personal, Indoor and Mobile Radio Communications, Sept. 2004. Barcelona, Spain.
18. G. Janssen, P.A. Stigter and R. Prasad, "Wideband indoor channel measurements and BER analysis of frequency selective multipath channels at 2.4, 4.75, and 11.5 GHz," IEEE Transactions on Communications, vol. 44, no. 10, October 1996, pp. 1272–1288.
19. J. Keignart and N. Daniele, "Subnanosecond UWB channel sounding in frequency and temporal domain", Ultra Wideband Systems and Technologies, Digest of Papers. IEEE Conference on, pp. 25–30, 2002
20. D. Cassioli, M.Z. Win and A.R. Molisch, "A statistical model for the UWB indoor channel," Vehicular Technology Conference, Spring, 2001, pp. 1159–1163.
21. R.C. Qiu and I.T. Lu, "Multipath resolving with frequency dependence for wide-band wireless channel modeling," IEEE Transactions on Vehicular Technology, vol. 48, no.1, January 1999, pp. 273–285.
22. R.J. Foerster, "Channel modelling sub-committee report," Technical report, P802. 15-02/490rl-SG3a, IEEE 802.15 Working Group for Wireless Personal Area Networks, February 2003.
23. K. Pahlavan and A.H. Levesque, Wireless Information Networks, John Wiley, NY, 1995
24. S.J. Howard and K. Pahlavan, "Autoregressive modeling of wide-band indoor radio propagation," IEEE Transactions on Communications, vol. 40, no. 9, September 1992, pp. 1540–1552.
25. A.F. Molisch et al., "IEEE 802.15.4a Channel model, final report," available at: www. ieee802.org/15/pub/TG4a.html
26. Z. Irahhauten, H. Nikookar and G. Janssen, "An overview of ultra wide band indoor channel measurements and modeling," IEEE Microwave and Wireless Components Letters, vol. 14, no. 8, pp. 386–388, August 2004, pp. 386–388.

Chapter 5
UWB Interference

Contents

In this chapter we study the interference issue of the UWB system. Regarding the interference two important, aspects should be noted, (i) the interference caused by the narrowband and wideband systems on the victim UWB system and (ii) the interference caused by UWB systems on the victim narrowband and wideband systems. Both interferences are important and should be considered in the design, evaluation and implementation of the systems. In Fig. 5.1 the spectrum of UWB system and other wireless systems are shown. As seen from this figure, several other services exist in or in the neighborhood of the UWB band. For example, the IEEE 802.11-a which works at 5.2 GHz is a main source of interference to indoor UWB systems. Other systems such as 2.4 GHz band WLANs as well as the GPS system (at 1.5 GHz), mobile cellular system (at 800 MHz and 1800 MHz band) are also source of interference to UWB systems. Mutually, UWB systems may affect existing wireless or navigation systems and cause interference to systems such as 802.11a, wireless systems in ISM band, Mobile cellular, and GPS.

The focus of this chapter is to study the effects of interference to and from UWB systems. First in Section 5.1 we provide an example of IEEE802.11-a interference on the UWB system where we evaluate numerically the effect of this source of interference at different distances on the UWB system. The general interference scenario of UWB system and the method of calculation

H. Nikookar, R. Prasad, *Introduction to Ultra Wideband for Wireless Communications*, Signals and Communication Technology,
DOI 10.1007/978-1-4020-6633-7_5,

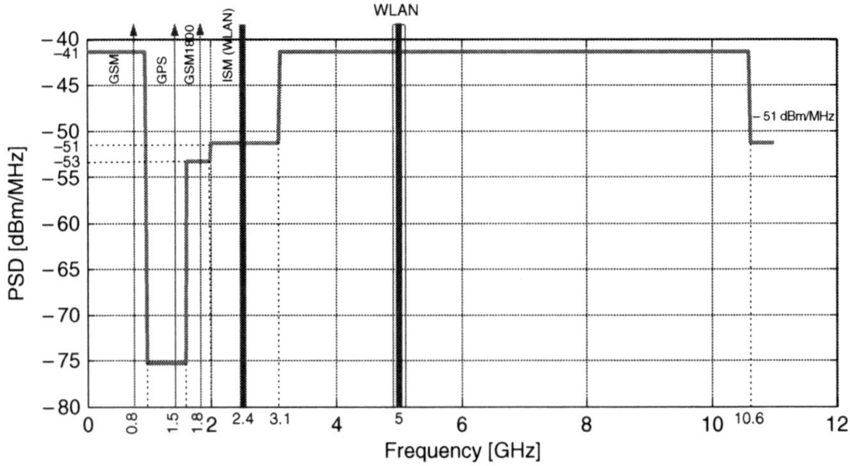

Fig. 5.1 UWB and other wireless systems spectrum

of signal-to-noise and interference ratio for the UWB receiver as well as for the general receiver is discussed in Section 5.2. Interference of UWB on a victim OFDM system is analyzed in Section 5.3 where the bit error results of OFDM system is sketched as a function of number of UWB interferers. In Section 5.4 the influence of UWB interference on the narrowband system is explained and the performance of narrowband victim receiver in the presence of UWB interferer will be derived. The coexistence of UWB and WiMax systems is studied in Section 5.5 and the effect of UWB interference on the Wimax receiver is evaluated. In Section 5.6 interference reduction methods are described and particularly in Section 5.7 the mitigation of interference of wideband systems on the UWB system using multicarrier templates is concentrated.

5.1 An Example: IEEE802-11.a Interference

To understand the effect of wideband interference on the UWB system, as example, an IEEE802.11-a interference source stationed at 5.25 GHz with a bandwidth of 200 MHz, is considered. The interference setup is shown in Fig. 5.2 and the normalized Power spectral density (PSD) values of the transmitted waveform and the interferer are illustrated in Fig. 5.3. The channel is assumed to be additive white Gaussian noise (AWGN). To gauge the propagation loss and the effect of interference, Friis transmission formula in free space is used:

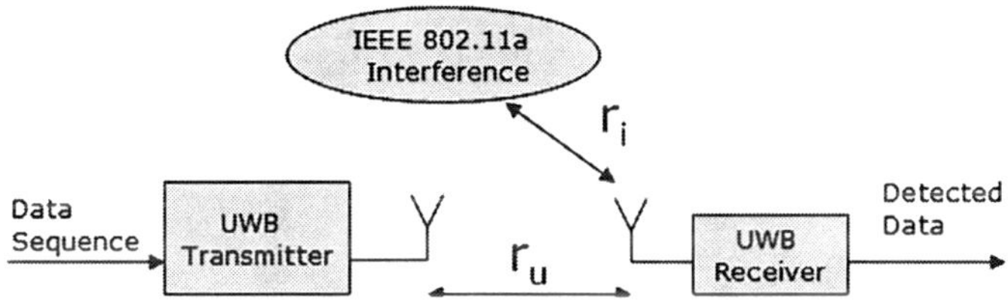

Fig. 5.2 Block diagram of the UWB system with IEEE802.11-a interferer

$$\frac{P_{\text{desired}}}{P_{\text{interf.}}} = \frac{P_{t\text{UWB}}}{P_{t_i}} \frac{\left(\frac{\lambda_{\text{UWB}}}{r_U}\right)^2}{\left(\frac{\lambda_i}{r_i}\right)^2} \tag{5.1}$$

where, $P_{\text{desired}}/P_{\text{interf.}}$ gives the ratio of desired signal power to the interferer power, $P_{t\text{UWB}}$ is the UWB transmission power based on the FCC emission limit. P_{ti} is the transmission power of the interferer (IEEE 802.11a) available from the specifications of these systems [1]. The UWB wavelength λ_{UWB} is obtained from the geometrical mean between UWB highest frequency and lowest frequency and λ_i is the interferer wavelength calculated from the center frequency of interferer. Parameters r_u and r_i are the distances between UWB transmitter to UWB receiver and interferer to UWB receiver, respectively. The Equation (5.1) can be re-written as:

$$\frac{r_i}{r_U} = \sqrt{\frac{P_{\text{desired}}}{P_{\text{interf.}}} \frac{P_{t_i}}{P_{t\text{UWB}}} \frac{\lambda_i}{\lambda_{\text{UWB}}}} \tag{5.2}$$

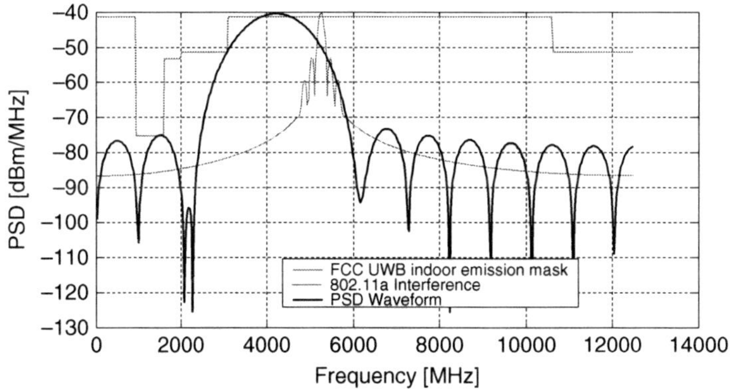

Fig. 5.3 PSD of transmitted UWB signal with 802.11a interference. The PSD curve of IEEE 802.11a has been normalized to the level of the FCC mask and the transmitted UWB waveform

Using this expression and for $P_{desired}/P_{interf.} = 0$ dB, $P_{tUWB} = 1$ mW, $P_{ti} = 100$ mW and $r_U = 1$ m, the value of r_i is obtained as 10 m. If $P_{desired}/P_{interf.} = 10$ dB, then the interfere distance r_i increases to 30 m. Which means that if 10 times larger power is desired at the UWB receiver side, assuming the same interferer power, the distance between interferer and UWB receiver should be 3 times larger.

5.2 General Method of Signal to Interference Ratio Calculation

In the previous section a simple case of only one interferer was considered. Moreover, the propagation channel considered was a simplistic free space loss model. In the real life, the situation is more complex. More interferers may be present and interfering with the UWB signal. The channel between each transmit and each receive point might be more complex as well. A block diagram of a general interference scenario is shown in Fig. 5.4. As seen from this figure a UWB communication link is between UWB transmitter and receiver. A general transmitter/receiver pair is in the neighborhood of the UWB system. Other

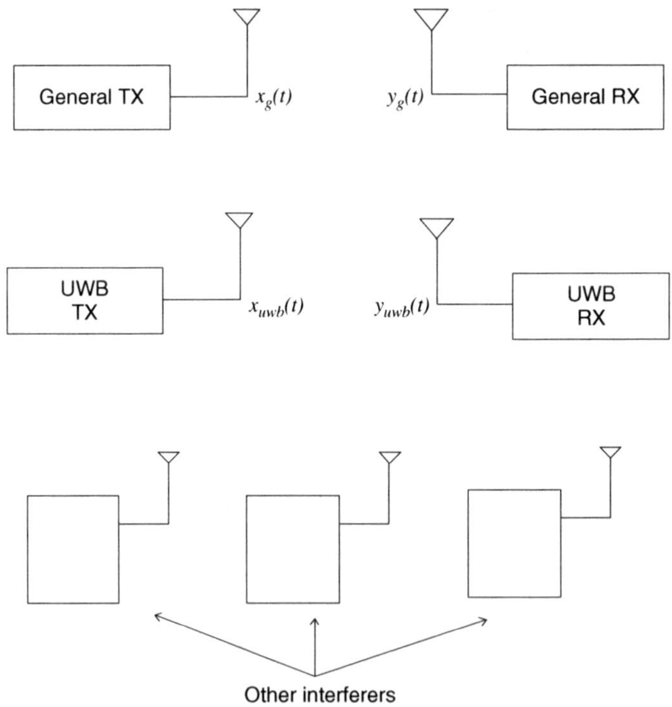

Fig. 5.4 General interference scenario

interferers are also present. We would like to evaluate the signal to interference and noise ratio for the UWB and general receiver. Let us consider the following notations:

$h_{uu}(t)$: Impulse response of channel between UWB TX and UWB RX
$h_{gu}(t)$: Impulse response of channel between General TX and UWB RX
$h_{ug}(t)$: Impulse response of channel between UWB TX and General RX
i_u(t): all other interferences at the input of UWB RX
$i_g(t)$: all other interferences at the input of General RX
$n_u(t)$: Equivalent receiver noise representing noise generated within UWB RX
$n_g(t)$: Equivalent receiver noise representing noise generated within General RX.

The signal at the UWB receiver can be obtained as

$$y_{uwb}(t) = h_{uu}(t)*x_{uwb}(t) + h_{gu}(t)*x_g(t) + i_u(t) + n_u(t) \tag{5.3}$$

and the signal at the general receiver is:

$$y_g(t) = h_{gg}(t)*x_g(t) + h_{ug}(t)*x_{uwb}(t) + i_g(t) + n_g(t) \tag{5.4}$$

Using the Fourier transform of any channel impulse response $h(t)$ as

$$H(f) = F\{h(t)\} = \int_{-\infty}^{\infty} h(t)e^{-j2\pi ft}\, dt \tag{5.5}$$

the power spectral density of the received signal at the UWB receiver (i.e., $S_{y\mathrm{UWB}}$) and general receiver (i.e., S_{yg}) are written, respectively, as [2]:

$$\begin{aligned} S_{y_{\mathrm{uwb}}} &= |H_{uu}(f)|^2 S_{x_{\mathrm{uwb}}} + |H_{gu}(f)|^2 S_{x_g} + S_{i_u}(f) + N_u \\ S_{y_g} &= |H_{gg}(f)|^2 S_{x_g} + |H_{ug}(f)|^2 S_{x_{\mathrm{uwb}}} + S_{i_g}(f) + N_g \end{aligned} \tag{5.6}$$

where, Nu is the spectral density height of noise generated in the UWB receiver, and Ng is the spectral density height of noise generated in the General receiver. The power of power of narrowband signal received by the general receiver is obtained as:

$$P_g = \int_0^{\infty} |H_{gg}(f)|^2 S_{x_g}(f)\mathrm{d}f \tag{5.7}$$

and Power of the UWB signal received by the general receiver

$$U_g = \left|H_{ug}(f_g)\right|^2 S_{X_{uwb}}(f_g) B_g \tag{5.8}$$

where *Bg* is the bandwidth of the general receiver. The power of Interference in the general receiver is

$$I_g = \int_{f_0 - B_g/2}^{f_0 + B_g/2} S_{i_g}(f)\mathrm{d}f \tag{5.9}$$

where f_0 is the center frequency of the general receiver. Now using Equations (5.7)–(5.9), the ratio of signal power to total interference and noise power for the general receiver (γ_g) is obtained:

$$\gamma_g = \frac{P_g}{U_g + N_g B_g + I_g} \tag{5.10}$$

The above procedure can be repeated for the calculation of signal-to-interference ratio at the UWB receiver.

The power of the UWB signal received by the UWB receiver is

$$P_U = \int_0^\infty |H_{uu}(f)|^2 S_{X_{uwb}}(f)\mathrm{d}f \tag{5.11}$$

The power of the narrowband signal received by the UWB receiver is obtained as

$$G_u = \int_0^\infty \left|H_{gu}(f_g)\right|^2 S_{X_g}(f)\mathrm{d}f \tag{5.12}$$

And the power of interference in the UWB receiver is

$$I_U = \int_0^\infty S_{i_u}(f)\mathrm{d}f \tag{5.13}$$

Using (5.11)–(5.13) the ratio of signal power to interference and noise power for the UWB receiver (γ_{UWB}) is calculated as:

$$\gamma_{\mathrm{UWB}} = \frac{P_U}{G_u + N_u B_{\mathrm{UWB}} + I_U} \tag{5.14}$$

Where B_{UWB} is the bandwidth of UWB system. According to this method knowing the channel impulse response between each transmit and receive end and knowing the spectral densities of noise and interferer signals, the signal- to-noise and interference power at the point of UWB receiver or general receiver can be calculated. The former is used to evaluate the effect of other interferers on the victim UWB system, and the latter is used to calculate the effect of UWB system on the general victim receiver working in the environment of UWB system.

5.3 Interference of UWB to Existing OFDM System

In Section 5.1 we studied the interference of IEEE 802.11 a on the UWB system and in the previous section the general method of signal-to-interference ratio (SIR) calculation was provided. Now we focus our attention on the effect of UWB system on a victim OFDM system. Our goal is to evaluate the bit error rate (BER) performance of the victim OFDM system when interfered by a UWB system in its neighborhood. We assume that the center frequency of OFDM system as well as its bandwidth falls in the band of UWB (i.e., 3.1–10.6 GHz). Let us consider the generic OFDM signal as:

$$s(t) = \sum_{m=0}^{M-1} \sum_{i=-\infty}^{\infty} b_m(i)\, e^{j2\pi f_m(t-iT)} \quad p(t-iT) \tag{5.15}$$

where $b_m(i)$ is the symbol of the mth subchannel at time interval iT, and for BPSK and QPSK modulation is ± 1 and $\pm 1 \pm j$, respectively, and $p(t)$ is the response of the transmitter filter which is a rectangular pulse with duration T and amplitude 1. The frequency of mth OFDM subcarrier is f_m

$$f_m = f_0 + \frac{m}{T} \qquad m = 0, 1, 2, \ldots, M-1 \tag{5.16}$$

where f_0 is the lowest frequency of the carriers, M is the number of subcarriers, and T is the OFDM symbol duration. Referring to (5.16) it is seen that subcarrier frequencies are separated by multiples of $1/T$. Meanwhile, in the OFDM system the carriers are orthogonal, i.e.,

$$\frac{1}{T}\int_0^T e^{j2\pi f_{m_1} t} e^{-j2\pi f_{m_2} t}\, dt = \begin{cases} 1 & if \;\; m_1 = m_2 \\ 0 & \text{otherwise} \end{cases} \tag{5.17}$$

Therefore, considering ideal transmission, and in spite of overlapping of the spectra, detection of the signal on one subchannel does not suffer from any other subchannels.

By transmission ofthe OFDM signal of (5.15) through the channel, with impulse response of $h(t)$

$$h(t) = \sum_{k=0}^{N-1} a_k \delta(t - \tau_k) e^{j\theta_k} \tag{5.18}$$

where a_k, τ_k , θ_k and N are respectively, amplitude, time, phase of arrival, and number of path components of the impulse response of the channel. The received signal $r(t)$ (i.e., the output of the channel), is obtained as:

$$r(t) = s(t)^* h(t) + n(t) \tag{5.19}$$

where * denotes the convolution and $n(t)$ is additive white Gaussian noise with the spectral density height of $N_0/2$. The first term of (5.19) is written as

$$s(t)^* h(t) = \sum_{k=0}^{N-1} \sum_{m=0}^{M-1} \sum_{j=-\infty}^{\infty} a_k \, b_m(i) e^{j[2\pi f_m(t-\tau_k-iT)+\theta_k]} \, p(t - iT - \tau_k) \tag{5.20}$$

In the receiver the recovery of data associated with the carrier f_k is performed by taking the decision variable z_k as

$$z_k = \int_0^T r(t) p(t) e^{-j2\pi f_k t} \, dt \tag{5.21}$$

which can be written as

$$z_k = \sum_{n'=0}^{N-1} \sum_{m=0}^{M-1} \sum_{i=-\infty}^{\infty} a_{n'} \, b_m(i) e^{-j[2\pi f_m(iT+\tau_{n'})-\theta_{n'}]} \int_0^T e^{j\frac{2\pi}{T}(m-k)t} \, p(t - iT - \tau_{n'}) \, p(t) \, dt$$
$$+ \int_0^T n(t) p(t) e^{-j\frac{2\pi k}{T}t} \, dt \tag{5.22}$$

or

$$z_k = \sum_{m=0}^{M-1} \sum_{n'=0}^{N-1} a_{n'} \left\{ b_m(-1) \int_0^{\tau_{n'}} e^{j[2\pi(f_m - f_k)t - \phi_{m,n'}]} \, dt + b_m(0) \int_{\tau_{n'}}^{T} e^{j[2\pi(f_m - f_k)t - \phi_{m,n'}]} \, dt \right\}$$
$$+ \int_0^T n(t) e^{j2\pi f_k t} \, dt \tag{5.23}$$

where $\varphi_{m,n'} = 2\pi f_m \tau_{n'} - \theta_{n'}$. Equation (5.23) can be written as

$$z_k = \sum_{n'=0}^{N-1} a_{n'} \left\{ b_k(-1) \int_0^{\tau_{n'}} e^{j\phi_{m,n'}} \, dt + b_k(0) \int_{\tau_{n'}}^{T} e^{j\phi_{m,n'}} \, dt \right\}$$
$$+ \sum_{m=0_{m\neq k}}^{M-1} \sum_{n'=0}^{N-1} a_{n'} \left\{ b_m(-1) \int_0^{\tau_{n'}} e^{j[2\pi(f_m-f_k)t-\phi_{m,n'}]} \, dt + b_m(0) \int_{\tau_{n'}}^{T} e^{j[2\pi(f_m-f_k)t-\phi_{m,n'}]} \, dt \right\}$$
$$+ \int_0^T n(t) e^{j2\pi f_k t} \, dt \tag{5.24}$$

Without loss of generality we can assume that receiver is matched to the first path of the multipath received signal, i.e., we assume $\tau_0 = 0$ and $\theta_0 = 0$. Therefore,

$$z_k = a_0 b_k(0) T + \sum_{n'=1}^{N-1} a_{n'} \left[b_k(-1) \int_0^{\tau_{n'}} e^{j\phi_{k,n'}} \, dt + b_k(0) \int_{\tau_{n'}}^{T} e^{j\phi_{k,n'}} \, dt \right]$$
$$+ \sum_{n'=0}^{N-1} \sum_{m=0, m\neq k}^{M-1} a_n \left\{ b_m(-1) \int_0^{\tau_{n'}} e^{j[2\pi(f_m-f_k)t-\phi_{m,n'}]} \, dt + b_m(0) \int_{\tau_{n'}}^{T} e^{j[2\pi(f_m-f_k)t-\phi_{m,n'}]} \, dt \right\}$$
$$+ \int_0^T n(t) e^{j2\pi f_k t} \, dt \tag{5.25}$$

where $b_m(-1)$ and $b_m(0)$ indicate the previous and current symbols respectively, which are transmitted at the carrier f_m. Equation (5.25) can be written as the following

$$z_k = a_0 b_k(0) T + \sum_{n=1}^{N-1} a_n \left[b_k(-1) X_{k,k}^n + b_k(0) \hat{X}_{k,k}^n \right]$$
$$+ \sum_{n=0}^{N-1} \sum_{m=0, m\neq k}^{M-1} a_n \left\{ b_m(-1) \left[X_{m,k}^n + Y_{m,k}^n \right] + b_m(0) \left[\hat{X}_{m,k}^n + \hat{Y}_{m,k}^n \right] \right\} + w_k \tag{5.26}$$

where

$$w_k = \int_0^T n(t) e^{j\frac{2\pi k t}{T}}\, dt \tag{5.27}$$

and

$$\begin{aligned} X^n_{k,k} &= \cos\phi_{k,n}\ R_{k,k}(\tau_n) & \hat{X}^n_{k,k} &= \cos\phi_{k,n}\ \hat{R}_{k,k}(\tau_n) \\ X^n_{m,k} &= \cos\phi_{m,n}\ R_{m,k}(\tau_n) & \hat{X}^n_{m,k} &= \cos\phi_{m,n}\ \hat{R}_{m,k}(\tau_n) \\ Y^n_{m,k} &= \sin\phi_{m,n}\ R'_{m,k}(\tau_n) & \hat{Y}^n_{m,k} &= \sin\phi_{m,n}\ \hat{R}'_{m,k}(\tau_n) \end{aligned} \tag{5.28}$$

and partial cross correlations are given by

$$R_{k,k}(\tau) = \int_0^\tau \mathrm{d}t = \tau$$

$$\hat{R}_{k,\,k}(\tau) = \int_\tau^T \mathrm{d}t = T - \tau$$

$$\begin{aligned} R_{m,k}(\tau) &= \int_0^\tau \cos\frac{2\pi(m-k)t}{T}\,\mathrm{d}t = T\,\frac{\sin\frac{2\pi(m-k)\tau}{T}}{2\pi(m-k)} \quad m \neq k \\ R'_{m,k}(\tau) &= \int_0^\tau \sin\frac{2\pi(m-k)t}{T}\,\mathrm{d}t = T\,\frac{1-\cos\frac{2\pi(m-k)\tau}{T}}{2\pi(m-k)} \quad m \neq k \\ \hat{R}_{m,k}(\tau) &= \int_\tau^T \cos\frac{2\pi(m-k)t}{T}\,\mathrm{d}t = -T\,\frac{\sin\frac{2\pi(m-k)\tau}{T}}{2\pi(m-k)} \quad m \neq k \\ \hat{R}'_{m,k}(\tau) &= \int_\tau^T \sin\frac{2\pi(m-k)t}{T}\,\mathrm{d}t = -T\,\frac{1-\cos\frac{2\pi(m-k)\tau}{T}}{2\pi(m-k)} \quad m \neq k \end{aligned} \tag{5.29}$$

From (5.26) it is seen that the first term is the desired signal, second term is due to intersymbol interference (ISI) caused by the multipath channel, third term relates to the loss of orthogonality between subcarriers also due to multipath fading of the wireless channel, which is intercarrier interference (ICI), and the last term due to noise. Accordingly, (5.26) can be rewritten as

$$z_k = \text{desired signal} + \text{ISI} + \text{ICI} + w_k \tag{5.30}$$

Where

$$\text{desired signal} = a_0 b_k(0) T \tag{5.31}$$

And

$$ISI = \sum_{n=1}^{N-1} a_n \left[b_k(-1) X_{k,k}^{n} + b_k(0) \hat{X}_{k,k}^{n} \right] \tag{5.32}$$

and

$$ICI = \sum_{n=0}^{N-1} \sum_{m=0_{m \neq k}}^{M-1} a_n \left\{ b_m(-1) \left[X_{m,k}^{n} + Y_{m,k}^{n} \right] + b_m(0) \left[\hat{X}_{m,k}^{n} + \hat{Y}_{m,k}^{n} \right] \right\} \tag{5.33}$$

Besides ISI, ICI and noise, the decision variable of (5.30) does have the effect of UWB interference U which can be considered as the total UWB interference impinged on the decision variable from I UWB users:

$$U = \sum_{i=1}^{I} U_i \tag{5.34}$$

where U_i is the interference contribution of the ith UWB user, and I is the total number of UWB interferers. Accordingly, (5.26) is re-written as:

$$z_k = \text{desired signal} + ISI + ICI + w_k + U \tag{5.35}$$

It is worth mentioning that one of the features of the OFDM technique is coping with the frequency selectivity of the channel. This is achieved by consideration of the *guard interval*. By exerting the guard period, the transmitted signal duration T is divided into two parts, i.e., a guard interval T_G and effective symbol duration of T', that is

$$T = T_G + T' \tag{5.36}$$

In this case the impulse response of the receiving filter will have a rectangular shape but with duration of T-T_G. By considering the guard period greater than the maximum excess delay of the multipath propagation channel (i.e., $T_G > \tau_{N-1}$ in Eq. (5.18)), the effect of ISI could be suppressed. That is, the period of integrals in (5.25) will start from $T_G > \tau_{N-1}$. Therefore, $X, \hat{X}, Y$ and $\hat{Y}$ in (5.32)–(5.33) will be zero and the decision variable can be simplified as:

$$z_k = \text{desired signal} + w_k + U \tag{5.37}$$

It should be mentioned that that by applying the guard interval in the OFDM signal, due to different durations for the transmitting and receiving filters, optimal matched filter condition is not completely fulfilled, and a fraction of the transmitted power is sacrificed in order to avoid ISI and ICI. Besides, this reduces the bandwidth efficiency by a factor of $10\log(1 + T_G/T')$ dB.

In order to evaluate the bit error rate performance of OFDM system in the presence of multipath, noise and other UWB interferers, we study the probability of error when the OFDM receiver makes decision on z_k. For the BPSK modulation the error probability is calculated as the real part of sampled signal of (5.37) be less than zero assuming a 1 has been transmitted or

$$P(E) = \text{Prob}(\text{Re}\{z_k\} < 0 | b_k(0) = 1) \tag{5.38}$$

Referring to (5.37) and by considering the interference terms (w_k and U) to be independent of each other and assuming that the decision variable can be approximated as a Gaussian random variable, the probability of error is calculated as

$$P(E) = Q\left\{\sqrt{\frac{a_0^2 T^2}{\text{Var}(z_k)}}\right\} \tag{5.39}$$

where $Q(.)$ is

$$Q(x) = \frac{1}{\sqrt{2\pi}} \int_x^{\infty} e^{-y^2/2} dy \tag{5.40}$$

The variance of the decision variable is written as

$$\text{var}(z_k) = \text{var}(w_k) + \text{var}(U) \tag{5.41}$$

The noise process is considered as zero mean Gaussian process. Using (5.27) and as discussed in Problem (5.5) its variance is obtained as

$$\text{var}(w_k) = \frac{N_0(T - T_G)}{2} \tag{5.42}$$

where $N_0/2$ is the spectral density height of the noise. The power of UWB interference can be calculated by the height of UWB power spectral density PSD_{UWB} (i.e., –41 dBm/MHz) at OFDM frequency (e.g., 5.2 GHz band of

IEEE 802.11-a), the bandwidth of OFDM system (B_{OFDM}) and the path loss rule. Assuming a path loss exponent of n_0 (see Chapter 4 Section 4.3) and a reference distance d_0, the total power of UWB interference at OFDM receiver can be obtained as:

$$Var(U) = I.(PSD_{UWB}B_{OFDM})\left(\frac{d}{d_0}\right)^{-n_0} \tag{5.43}$$

Where I is the number of independent UWB interferers and PSD_{UWB} is the power spectral density of each UWB interferer. It should be noted that as amplitude fading a_0 is a random variable the bit error probability of (5.39) has to be averaged over the probability density function of the fading, i.e., $f_A(a_0)$:

$$P(E) = \int_0^{\infty} Q\left(\sqrt{\frac{a_0^2 T^2}{Var(z_k)}}\right) f_A(a_0)da_0 \tag{5.44}$$

Using the above-mentioned analytical procedure, the bit error probability of OFDM system versus distance when different number of UWB interferences are heard at the OFDM receiver is sketched in Fig. 5.5. It is seen that the UWB interference can have a detrimental effect on the OFDM system if the interfering UWB systems are in the vicinity of the OFDM receiver. Obviously, as the number of UWB interferers increases the bit error performance degrades.

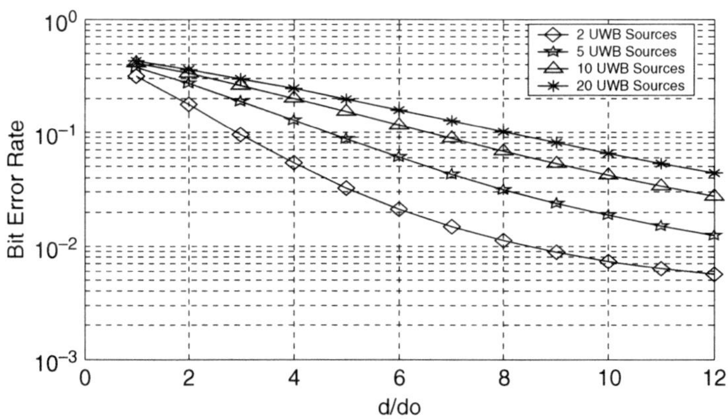

Fig. 5.5.a BER performance curves as a function of distance (d/do), for various number of UWB interference sources. The plots have been drawn considering parameters of the WLAN standard IEEE 802.11a. The received signal energy within a single pulse to noise spectral density height (Eb/No) has been taken to be 5 dB

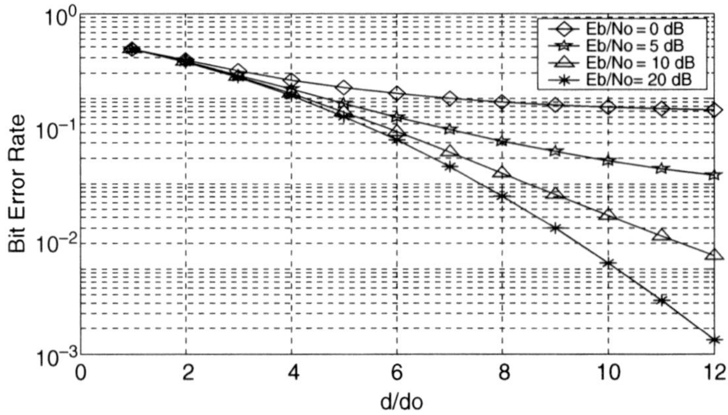

Fig. 5.5.b BER performance curves as a function of distance (d/d0), for various received signal energy within a single pulse to noise spectral density height (Eb/No). The plots have been drawn considering parameters of the WLAN standard IEEE 802.11a. And the number of UWB interferers considered is 5.

5.4 Interference of UWB to Narrowband Systems

Due to inherent nature of UWB systems in sharing the spectrum, it is important to study the effect of the UWB interference on a victim narrowband system. Consider the received signal at the input of a narrowband receiver as:

$$r(t) = s(t) + u(t) + n(t) \tag{5.45}$$

Where $s(t)$ is the narrowband signal expressed as:

$$s(t) = \sqrt{2P_s}\cos(2\pi f_0 t + \theta) \sum_{k=-\infty}^{\infty} b_k v(t - kT_0) \tag{5.46}$$

P_s is the average transmit power, f_0 is the center frequency, θ is a random phase, $b_k \in \{\pm 1\}$ is the modulated symbol of the narrowband transmitter, $v(t)$ is the transmitted narrowband waveform and T_0 is the symbol duration. The narrowband receiver is considered as a coherent matched filter with impulse response as:

$$h_{MF}(t) = 2\cos[2\pi f_0(T_0 - t) + \theta]v(T_0 - t) \tag{5.47}$$

In , $n(t)$ is the AWGN noise and $u(t)$ is the UWB signal. The sampled output of the matched filter at T_0 can be written as:

$$z = y_s + y_u + y_n \tag{5.48}$$

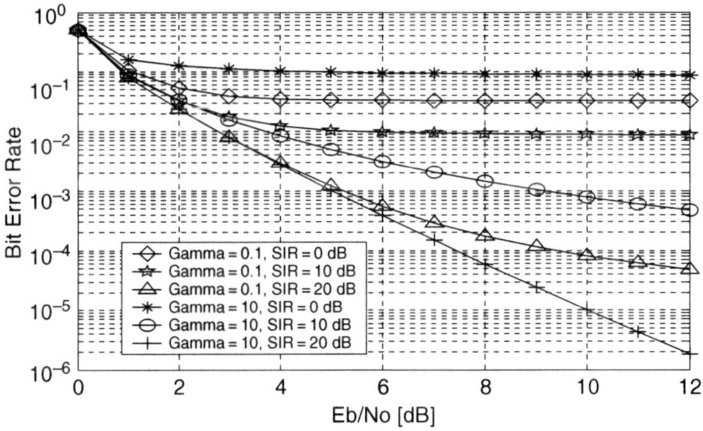

Fig. 5.6 Bit Error Rate curves for various values of γ and SIR

Where

$$\begin{aligned} y_s &= s(t) * h_{MF}(t)|_{T_0} \\ y_u &= u(t) * h_{MF}(t)|_{T_0} \\ y_n &= n(t) * h_{MF}(t)|_{T_0} \end{aligned} \tag{5.49}$$

y_s, y_u and y_n are respectively the sampled response of the matched filter to the desired narrowband signal, UWB signal, and noise. It can be shown that [3]:

$$y_s = \sqrt{2P_s} b_k T_0 \tag{5.50}$$

The mean of the noise term y_n is zero and its variance is:

$$\sigma_{y_n}{}^2 = N_0 T_0 \tag{5.51}$$

The UWB signal occupies a much larger bandwidth than the narrowband receiver. Therefore, its spectrum can be considered relatively constant in the bandwidth of the narrowband system. Therefore, the signal-to-interference ratio (SIR) within the bandwidth of the narrowband receiver can be obtained as:

$$SIR = P_s/(PSD_{UWB} B_{NB}) \tag{5.52}$$

Where PSD_{UWB} is the power spectral density of UWB (to be imposed by FCC) and B_{NB} is the bandwidth of narrowband receiver $= 1/T_0$. Let us denote the ratio of pulse repetition frequency (PRF) of UWB signal to the bandwidth of narrowband receiver as $\gamma = PRF/B_{\mathrm{NB}}$. When $\gamma >> 1$, the UWB interference can be approximated by Gaussian noise which provides a simple way for narrowband systems to determine the performance loss due to a UWB transmitter by looking at how much the UWB interference raises the noise floor of the narrowband receiver. With this approximation the probability of bit error is obtained as [3]:

$$P(E) = Q\left(\sqrt{\frac{2}{\frac{N_0}{E_b} + \frac{1}{SIR}}}\right) \tag{5.53}$$

Where E_b is the energy per bit, $N_0/2$ is the spectral density height of the noise and $Q(.)$ is expressed by (5.40). Usually the Gaussian approximation is accurate (within 1 dB) of actual performance when $\gamma > 5$. It is shown in [3] that for $\gamma < 1$ the error probability can be written as:

$$P(E) = \gamma Q\left(\sqrt{\frac{2}{\frac{N_0}{E_b} + \frac{1}{\gamma SIR}}}\right) + (1-\gamma)Q\left(\sqrt{\frac{2E_b}{N_0}}\right) \tag{5.54}$$

As shown in Fig. 5.6 as well as from the above explained analysis it is clear that the performance of narrowband victim receiver in the presence of UWB interferer depends on a number of parameters such as the PRF of UWB interferer with respect to the bandwidth of narrowband signal as well as the ratio of narrowband signal power to the UWB signal power within the bandwidth of the narrowband receiver.

5.5 Interference to WiMax

Another example of the effect of UWB interference is on the WiMax receiver. WiMax systems based on IEEE 802.16 standard, working in the frequency range of 2–11 GHz, provide high data rates (in the order of 75 Mbps) to a broad geographical area [4], [5]. UWB systems operating in the 3.1–10.6 GHz share the band with WiMax. Now we study the coexistence of these two systems. Consider an UWB system is working in the vicinity of the victim WiMax system. The UWB signal power at the location of WiMax system is expressed as:

$$P = \frac{\int_{f_l}^{f_h} S_{UWB}(f)df}{PL(r)} \tag{5.55}$$

Where S_{UWB} is the PSD of UWB signal, f_l and f_h are respectively the lowest and highest frequencies of the UWB band, and $PL(r)$ is the path loss as a function of distance r between UWB and WiMax systems. For the path loss let us consider a free space formula:

$$PL(r) = PL(r_0)\left(\frac{r}{r_0}\right)^2 \tag{5.56}$$

where $PL(r_0)$ is the path loss at reference distance r_0. If a set of I UWB systems working in a hot spot and distributed between $[r_{\min}, r_{\max}]$ in the vicinity of WiMax receiver, the total interference power at the victim receiver will be [6]:

$$P = K\sum_{i=1}^{I}\int_{f_l}^{f_h}\int_{r_{\min}}^{r_{\max}}\int_{\varphi=0}^{2\pi}\frac{S_{UWB,i}(f)}{r^2}\,dr\,d\varphi\,df \tag{5.57}$$

Where $S_{\mathrm{UWB},i}$ is the power spectral density of the ith UWB interferer, and $K = {r_0}^2/PL(r_0)$. Knowing the number of UWB interferers, and the min and max distances as well as reference distance and its path loss, the received power can be calculated and compared with the maximum tolerable interference for WiMax shown in Table 5.1. In [6] it is concluded that for different values of the WiMax receiver bandwidths and up to 50 UWB interferers, the total interference maintains below the threshold levels in Table 5.1.

Table 5.1 Maximum tolerable interference for WiMax

BW(MHz)	1.25	2.5	10	20
OFDM WiMax (dBW)	–134.5	–131.5	–127	–123.9

5.6 Interference Reduction

The effect of UWB interference on the wideband system (such as IEEE 802.11-a) can be mitigated by using a notch filter. The spectrum of the UWB interfere signal is filtered (around 5.2 GHz) and its spectral contents in this band is suppressed. Another method of mitigating the effect of UWB interference on the victim wideband systems is designing time-hopping codes in such a way that the power spectral density of the UWB signal has less power in the band of wideband system (e.g., IEEE 802-11-a). Multibanding is another way of mitigation of UWB interference on narrowband and wideband systems. According to this method, the UWB band is divided in subbands and the transmit subbands are selected according to an interference threshold [7].

The notch filtering of UWB signal is a well known method of UWB interference mitigation. The time hopping code design is an interesting method but can only be applied to impulse radio UWB systems. The multi-band method is a practical solution that can be combined with notch filtering method in each subband. It is scalable which means that the bit rate, power consumption, complexity and cost can scale with the demand. Multiband UWB can coexist with other spectrum users as sub-bands can be turned off to avoid interference [7].

The effect of narrowband interference on the victim UWB system can be simply reduced by using a rejection filter. The frequency band of the interferer should fall in the stopband of the filter. The filter can be inserted in the receiver before correlator (which correlates the received signal with the transmit template waveform). According to the center frequency of the interferer, the cutoff frequency, the ripple in the passbend and attenuation in the stopband can be designed. Use of the rejection filter improves the BER performance of the UWB system when narrowband interferer exists [8]. However, in the absence of the interferer and when the channel is only additive white Gaussian noise (AWGN), the rejection filter slightly reduces the BER performance.

5.7 Interference Mitigation of Wideband System on UWB Using Multicarrier Templates

As mentioned earlier in this chapter, performance degradation of the UWB communication system due to interference from other services is a major issue. To reduce the effect of interference, the usage of modified template in the UWB transceiver and its multi-carrier representation has recently been reported [8]–[9]. By using template waveforms or its multi-carrier type representation at the transceiver, the interference power can be to a large extent mitigated. In this section we study the performance deterioration due to co-existing sources on the UWB transmission as well as the effects of using template waveforms at the transceiver. The fundamental thrust is transmission waveform shaping. Based on the radio spectrum scenario, the system dynamically adapts the transmission spectrum through the use of multi-carrier templates to evade the harmful effects of the interferer. The study includes the impact of waveform representation with multi-carrier templates, at both transmitter and receiver ends on the system performance in the presence of wideband interference. Results show that this interference mitigation technique is effective, allowing for coexistence of UWB system with different wideband systems.

The procedure for UWB data transmission and reception is illustrated in Fig. 5.7. The major blocks in the transmitter are the radio spectrum estimator, transmission waveform shaper and modulator. A signal decorrelator, integrator and detector are the key elements of the receiver. At the transmitting end, first a data source generates an arbitrary stream of data derived from the source alphabet. The spectrum estimator senses the spectrum and detects the presence of interference regions. Based on the spectrum estimates, a pulse waveform that has little or no energy in the interference domains is constructed by a proper

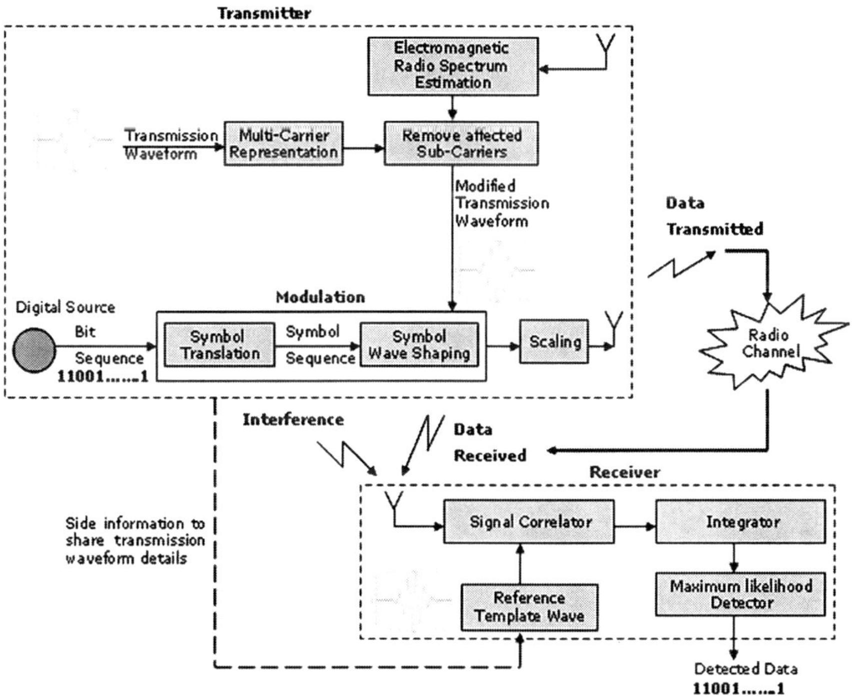

Fig. 5.7 System Model of the UWB transmission. The major blocks in the transmitted include the radio spectrum estimator, transmission waveform shaper and modulator. And at the receiver the main components are signal decorrelator, integrator and detector

choice of wavelet packets based codes. The stream of data is then linearly modulated by the pulse waveform to obtain a UWB signal. The UWB signal is then transmitted to the channel. At the receiver, the signal, corrupted by interference, is decorrelated with a similar wavelet packets based reference waveform. An integrator follows the correlator. And the data is detected using a using a Maximum likelihood detector. The transmitter constantly appraises the receiver on the transmission waveform nature by sending side information.

The transmitted signal $s(t)$ may be given as:

$$s(t) = g(t)\sin(2\pi f_o t) \tag{5.58}$$

where $g(t)$ is the pulse waveform, and f_o the centre frequency. Typically the $g(t)$ is taken to be a Gaussian waveform or its higher order derivatives. The modified template waveform is constructed by representing the transmission waveform as a multi-carrier-type template wave consisting of several sub-band pulses. When interferences exist, the modified template waveform is reconstructed by removing those sub-band pulses that are close to the interfering spectrum. In this way it is possible to reduce the effect of the interferers.

The multi-carrier template representation of the transmission waveform is given as:

$$s(t) \approx \sum_{n} W_{\psi}(nf_s)\psi\left(\frac{t}{nf_s}\right) \tag{5.59}$$

here nf_s is the frequency of the nth sub-carrier and $W_{\psi}(nf_s)$ are wavelet transform coefficients of $s(t)$ given by:

$$W_{\psi}(nf_s) = \int_{-\infty}^{\infty} \psi\left(\frac{t}{nf_s}\right)s(t)dt \tag{5.60}$$

$$\psi\left(\frac{t}{nf_s}\right) = g'(t)\sin(2\pi nf_s t) \tag{5.61}$$

where $g'(t)$ has the same form as $g(t)$ but with an adaptable pulse-width.

Based on the spectrum estimates available, those sub-carriers whose frequency components fall in the neighbourhood of the interference spectrum region are removed. The modified transmission waveform $\tilde{s}(t)$ is then reconstructed from the unaffected coefficients and is given by –

$$\tilde{s}(t) \cong \sum_{p} W_{\psi}(pf_s)\psi\left(\frac{t}{pf_s}\right), \; p \subset n, \notin m \tag{5.62}$$

where m denotes the set of sub-carriers that fall in the region of interference spectrum and p denotes the set of all sub-carriers other than m.

Now we discuss the influence of using modified template waveforms on enhancing system performance.

A. Impact of interference on UWB transmission – Performance of the UWB system under IEEE 802.11a interference.

Figure 5.8 shows the BER performance curves of the UWB system in the presence of a wideband IEEE 802.11a co-existing user for various $\frac{P_{\text{desired}}}{P_{\text{interference}}}$ ratios.

From the figures it is evident that the presence of IEEE 802.11a co-existing user affects the UWB system performance. The performance worsens with decreasing $\frac{P_{\text{desired}}}{P_{\text{interference}}}$ ratio. With decreasing $\frac{P_{\text{desired}}}{P_{\text{interference}}}$ ratio, the interference to the UWB channel increases and hence the deterioration in performance.

It is clear from the results that the presence of co-existing users can cause impairments to the extent that normal system operation becomes impossible. There is therefore the need for a mechanism that can enable the systems to

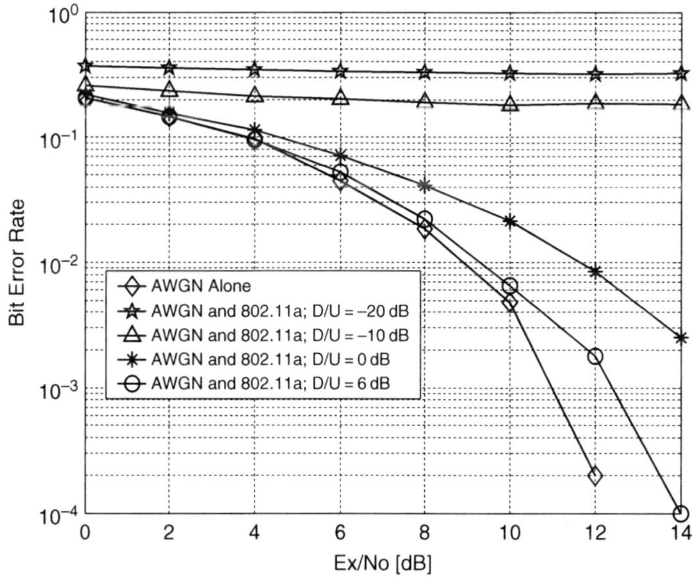

Fig. 5.8 BER performance of UWB in the presence of AWGN and IEEE 802.11a. In the figure the ratio $\frac{P_{\text{desired}}}{P_{\text{interference}}}$ is represented as D/U

coexist. The multi-carrier template wave technique proposed above is one such mechanism.

In the following set of results the $\frac{P_{\text{desired}}}{P_{\text{interference}}}$ ratio is taken to be −10 dB for the IEEE 802.11a interferer.

B. Influence of using modified template waveforms on enhancing system performance –
We shall now see the improvements brought by representing the UWB waveform by a multi-carrier template and by selectively removing carriers affected an IEEE 802.11a interferer. The BER performance curves of the UWB system in the presence of an IEEE 802.11a interferer is given in Fig. 5.9. Both the transmitter and receiver use multi-carrier type templates. The plots illustrate the improvements in BER performance brought about by using multi-carrier type template implementation at both the transmitter and receiver end and by selectively removing carriers in the neighbourhood of the interferer spectrum. From the plots, it is clear that the template representation of the UWB transmission waveforms does have a positive effect on the system performance.

As more and more carriers are removed from the neighborhood of the interferer, the energy in the region of the coexisting interference source is reduced and the BER performance of the system improves. The best performance is obtained when using the template wave with four sub-band pulses, centered at 4.6 GHz, 4.8 GHz, 5 GHZ and 5.2 GHz, respectively, removed. For

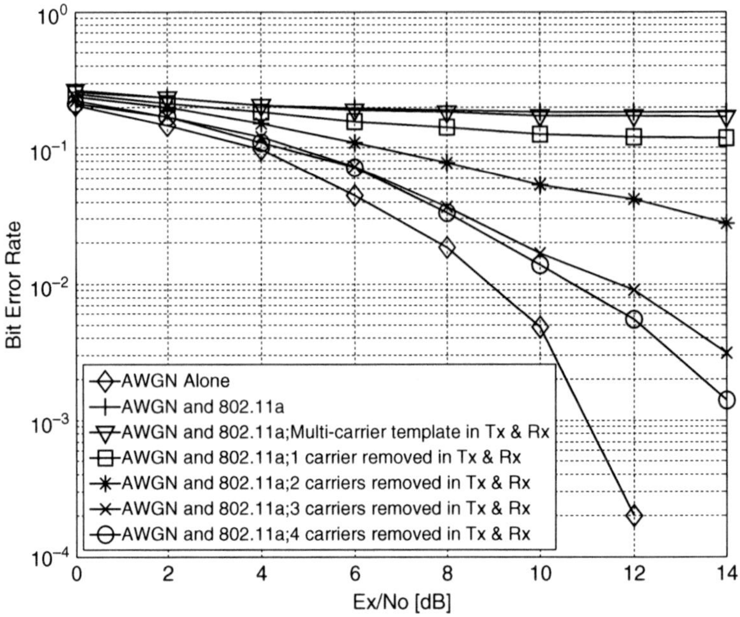

Fig. 5.9 BER curves for system under IEEE 802.11a interference while using multi-carrier template at both the transmitter and receiver ends. The notations Tx and Rx in the legend denote transmitter and receiver, respectively. The sub-carriers from the template waveform are removed as follows – 1 carrier centered at 5.2 GHz, 2 carriers centered at 5.0 GHz and 5.2 GHz, 3 carriers centered at 4.8 GHz , 5 GHz and 5.2 GHz, 4 carriers centered at 4.6 GHz, 4.8 GHz , 5 GHz and 5.2 GHz

this case, the system shows significant improvements in comparison to the ordinary template. For example, for a BER of 0.001 and less, the system under IEEE 802.11a interference performs only 3 dB worse in comparison to the AWGN case.

The modified transmission waveforms can be used at the transmitter alone or at the receiver alone (as the reference waveform) or at both the transmitter and receiver ends. The transceiver waveform combinations also influence the system performance in two ways:

1. Extent to which the transmitter-receiver template waveform combination neutralizes the corruption of the useful signal due to interference –
2. Sensitivity to loss of correlation between transmitted template waveform and the reference waveform when they are different

For the case of interference from IEEE 802.11a sources, it has been inferred that it is enough to use such a multi-carrier representation at the receiver end alone [10].

In this Section we studied a method to mitigate the impact of wideband (IEEE 802.11-a) interference on the performance of the victim UWB communication through use of multi-carrier type transmission pulses and template waveforms and by removing carriers from the interference region. Such representation improves performance when either the receiver end alone or both receiver and transmitter ends use multi-carrier type templates. It is worth mentioning that the above discussed method can be well used in the cognitive radio area, where adaptive sensing of the spectrum is performed and interference is mitigated accordingly.

5.8 Summary

The huge bandwidth of the UWB transmission means spans frequencies commonly used as carrier frequencies by other wireless services. Accordingly degradation in performance of UWB communication due to interference from co-existing services is a major issue. The focus of this chapter was to study the effects of interference to and from UWB systems. First the general method of signal to interference calculation was explained. Then, in addition to interference of UWB to narrowband systems, the effects of UWB interference on the BER performance of WLAN OFDM system as well as WiMax system was studied and analysed. Further in this chapter, the effect of narrowband and wideband interferences on the victim UWB system was studied and methods to mitigate the interference was discussed.

Problems

Problem 5.1 Consider the interference scenario explained in Section 5.1. But, instead of Friis law assume the path loss exponent n as:

$$n = \begin{cases} 1.0 & \text{if } r < 10\,\text{m} \\ 4.0 & \text{if } r \geq 10\,\text{m} \end{cases}$$

Using the expression (5.2) and for $P_{\text{desired}}/P_{\text{interf.}} = 0$ dB, $P_{tUWB} = 1$ mW, $P_{ti} = 100$ mW and $r_U = 1$ m, calculate the value of r_i. If $P_{\text{desired}}/P_{\text{interf.}} = 10$ dB, what will be the interfere distance r_i?

Problem 5.2 Repeat Problem 5.1 if the path loss model is expressed by:

$$PL \propto \left(\frac{d}{d_0}\right)^n$$

where d_0 is the reference distance and path loss exponent n is a function of frequency as:

$$n = n_0 + K(f - 5\,GHz)$$

n_0 is the free space path loss exponent ($n_0 = 2$), and K is a constant ($K = 1.58 \times 10^{-9}$). Compare the result with the results of Section 5.1. What do you conclude?

Problem 5.3 In Section 5.3 we studied in detail the bit error probability of OFDM system when I UWB interferers affecting the OFDM victim receiver. Now suppose each OFDM branch has a power of P_m. The total transmit power $P_{tot} = P_0 + P_1 + \ldots + P_{M-1} = 1$. The transmitted OFDM signal is:

$$x(t) = \sum_{m=0}^{M-1} \sqrt{P_m} b_m e^{j2\pi f_m t}$$

where M is the number of subcarriers, f_m is the mth OFDM subcarrier and b_m is the symbol of the mth sub-channel. Using the method explained in Section 5.3 and assuming a guard interval larger than the maximum delay of the channel, and $I > 5$ independent UWB interferers, calculate the average probability of bit error if

$$P_m = \frac{A}{s\sqrt{2\pi}} e^{\frac{\left(m - \frac{M}{2}\right)^2}{s^2}}$$

where A is a constant to have $P_{tot} = 1$, and $s = M/4$.
Repeat the problem if $s = M/16$ and $M/32$. What do you conclude?

Problem 5.4 Using the frequency dependent path loss model described in Problem 5.2 calculate the total interference power due to I independent UWB interferers distributed between $[r_{min}, r_{max}]$ in the vicinity of WiMax receiver. For $r_0 = 1$ m, $PL(r_0) = 57.6$ dB, $r_{min} = 2$ m, $r_{max} = 10$ m, and $I = 10$ UWB interferers, obtain the value of interference level and compare it with the maximum tolerable interference level for WiMax system.

Problem 5.5 Starting from (5.27) and assuming a zero mean Gaussian noise with two sided spectral density height of $No/2$, and a Guard interval of T_G for OFDM, obtain the variance of noise term (5.42).

Problem 5.6 In this problem we want to study the effect of UWB interference together with the impulsive noise on the performance of a victim OFDM receiver. The block diagram of the system is shown in Fig. P.5.6.1. The received signal at the OFDM receiver is corrupted by the interference from I UWB interferers and the impulsive noise. We consider a Bernoulli-Gaussian process to model the Impulsive Noise (IN). The random time of occurrence of the

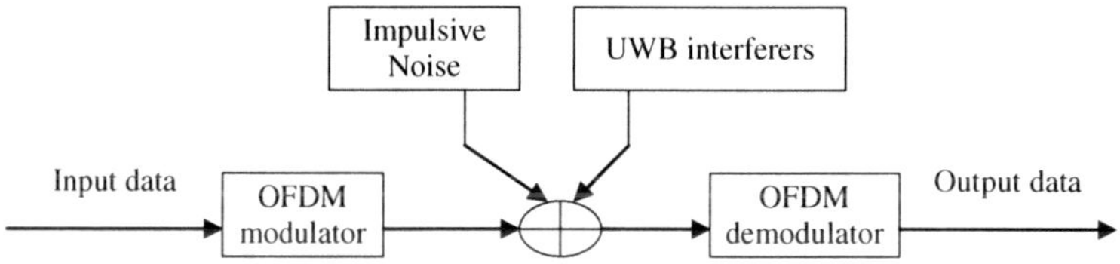

Fig. P.5.6.1 Block diagram of the OFDM system with UWB interferers and impulsive noise

impulses is modelled by a Bernoulli process $b(k)$, where K is the time point and $b(k)$ is a binary-valued process that takes a value of "1" with a probability of α and a value of "0" with probability of $1-\alpha$. The amplitude of the impulses is modelled by a Gaussian process $g(k)$ with mean zero and variance $2\sigma_g^2$. Each impulse is shaped by a filter with the impulse response $h(k)$, [11]–[12]. The Bernoulli-Gaussian model of impulsive noise is illustrated in Fig. P.5.6.2. The IN can be expressed as

$$n(k) = \sum_{i=0}^{P-1} h(i)g(k-i)b(k-i)$$

where P is the length of the impulse response of the impulse shaping filter. The probability density function (pdf) of impulsive noise $n(k)$ is given by

$$pdf_N^{BG}(n(k)) = (1-\alpha)\delta(n(k)) + \alpha pdf_N(n(k))$$

where $\delta(n(k))$ is the Kronecker delta function and $pdf_N(n(k)) = \frac{1}{\sqrt{2\sigma_g^2}\sqrt{2\pi}} e^{-\frac{1}{2}\left(\frac{n(k)}{\sqrt{2\sigma_g^2}}\right)^2}$ is the probability density function of a zero mean Gaussian process.

Using the Bernoulli-Gaussian model for the impulsive noise and considering an OFDM system with BPSK modulation, $M=32$ carriers and a guard time

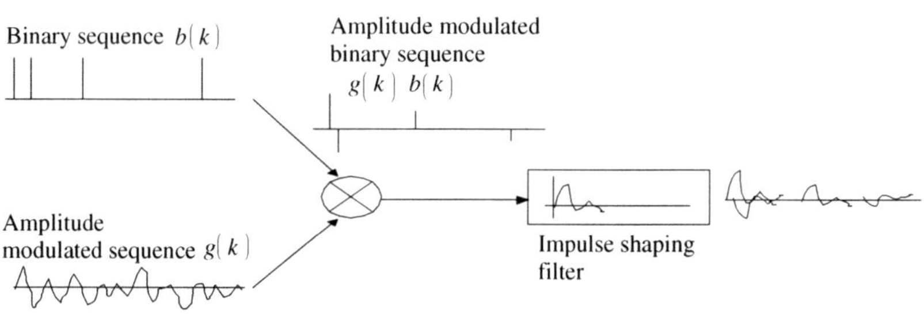

Fig. P.5.6.2 The impulsive noise modelled as the output of a filter excited by an amplitude-modulated binary sequence

larger than the maximum delay of the channel and assuming $I > 5$ UWB interferers, obtain the bit error probability of the OFDM transmission for $\alpha = 0.1$ and $\alpha = 0.01$.

References

1. 802.11-a IEEE Standard, Sept. 1999, Part 11: Wireless LAN medium access control and physical layer specifications: High speed physical layer in the 5 GHz.
2. R.A. Scholtz et al., "UWB Radio development challenges," IEEE MilCom, 2000, pp. 620–625.
3. J.R. Foerster, "Interference modelling of pulse-based UWB waveforms on narrowband systems," IEEE Vehicu, Technol. Conf., VTC-Spring 2002, pp. 1931–1935.
4. A. Ghosh et al., "Broadband wireless access with WiMax/802.16: Current performance benchmarks and future potential," IEEE Communication Magazine, 2005, vol. 43, no. 2, pp. 129–136.
5. 802.16-2004 IEEE standard for local metropolitan area networks, Part 16: Air interface for fixed broadband wireless access systems.
6. K. Sarfaraz et al., "Performance of Wi-Max receiver in presence of DS-UWB system," IEE Elect. Letters, vol. 41, no. 25, 8 December 2005.
7. S. Ghassemzadeh and V. Tarokh, "UWB Interference issues," June 2003, online: www.deas.harvard.edu/hbbcl/IMS_2003_int.pdf
8. T. Ikegami and K. Ohno, "Interference mitigation study for UWB radio", Proc. of the 14th IEEE International Symposium on Personal, Indoor and Mobile Radio (PIMRC), pp. 583–587, October 2003.
9. K. Ohno, T. Ikebe and T. Ikegami, "A proposal for an interference mitigation technique facilitating the coexistence of bi-phase UWB and other wideband systems," Proc. Internal Workshop on Ultra Wideband Systems Joint with Conference on Ultra Wideband Systems and Technologies (Joint UWBST & IWUWBS 2004), WA2-5, 2004.
10. M. Lakshmanan and H. Nikookar, "UWB Interference mitigation using multi-carrier templates," European Conference Wireless Technology, September 2006, Manchester, UK, pp. 115–118.
11. H. Nikookar and D. Nathoeni, "Performance evaluation of OFDM transmission over impulsive noisy channels," IEEE PIMRC Conference 2002, pp. 550–554.
12. S. Vaseghi, "Advanced Digital signal Processing and Noise Reduction" 2nd edition, John Wiley and sons NY, 2000.

Chapter 6
UWB Signal Processing

Contents

In this chapter we study the modulation schemes of impulse radio IR-UWB. This includes the UWB data and multiple access modulations. The bit error rate (BER) performance of M-ary pulse amplitude modulation and Pulse position modulation is discussed and compared. Then we will study different reception techniques, such as Rake and Transmit-Reference receivers. We will also evaluate the range-data rate performance of IR-UWB and the tradeoff between range and speed of transmission. Finally, we will assess the UWB capacity and the minimum signal-to- noise ratio required in the UWB receiver to reach the Shannon capacity limit.

6.1 Modulation

6.1.1 Data Modulation

Data in the IR-UWB communication systems can be modulated using different modulation schemes. The first one is the bi-phase modulation (BPM) wherein the data is encoded in the polarity of the impulses. See Fig. 6.1-a. The BPM signal can be expressed as:

$$s(t) = \sum_{k=-\infty}^{\infty} a_k \, p(t - kT_f), \text{ with } a_k = 1, -1 \tag{6.1}$$

H. Nikookar, R. Prasad, *Introduction to Ultra Wideband for Wireless Communications*, Signals and Communication Technology,
DOI 10.1007/978-1-4020-6633-7_6, © Springer Science+Business Media B.V. 2009

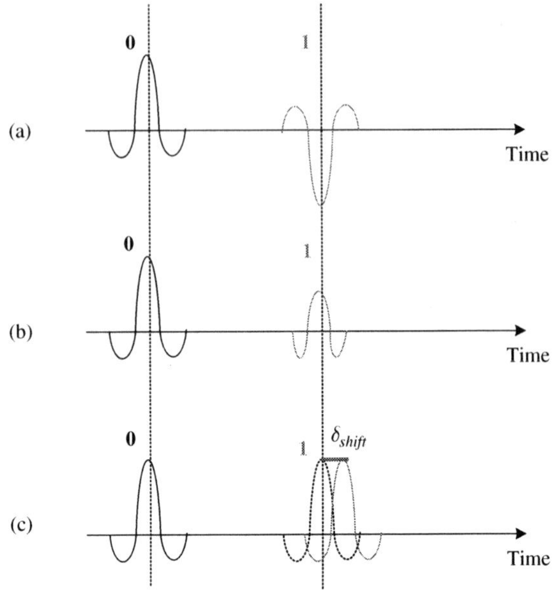

Fig. 6.1 Illustration of (**a**) Bi-Phase modulation, (**b**) Pulse amplitude modulation and (**c**) Pulse position modulation

where $p(t)$ is the pulse shape and T_f is the duration of time frame.

The second data modulation scheme is pulse amplitude modulation (PAM), which is based on encoding the data in the amplitude of the impulses. The PAM modulated signal can be written as:

$$s(t) = \sum_{k=-\infty}^{\infty} a_k.p(t - kT_f) \tag{6.2}$$

where a_k is the amplitude of the pulse $p(t)$ and T_f is the frame duration. As can be seen from Fig. (6.1-b) and Eq. (6.1), BPM is a form of PAM modulation. Specifically, when in the binary PAM modulation a_ks have symetric antipodal amplitudes this modulation reverts to the BPM modulation.

On-Off Keying (OOK) modulation is the simplest form of pulse modulation, in which the transmission of a pulse represents a data bit "1" and its absence represents a data bit "0". The general signal model for OOK is given by:

$$s(t) = \sum_{k=0}^{+\infty} a_k\, p(t - kT_f), \text{ with } a_k = 1, 0 \tag{6.3}$$

where again a_k is the amplitude of the pulse which can have the values 1 or 0 and T_f is the frame duration. The most preferable used modulation is the pulse position modulation (PPM) because it smoothes the spectrum of the UWB signal. The data is encoded by adding an extra time shift "δ_{shift}" to the impulse as shown in Fig. 6.1-c. The binary PPM signal is given by:

$$s(t) = \sum_{k=-\infty}^{\infty} p(t - kT_f \pm \delta_{shift}) \tag{6.4}$$

where the data modulation is done by small shifts in the pulse position δ_{shift}, $p(t)$ is the UWB pulse and T_f is the frame duration.

In the UWB transmission pulses are sent at regular intervals (i.e., pulse repetition interval-PRI). Because of this pulse repetition, peaks appear in the spectrum of the transmitted pulses. These peaks are generally referred to as "spectral lines" or "comb lines", as shown in Fig. 6.2. These spectral lines are undesirable as they limit the total transmit power. They are also not favorable regarding the FCC mask and the interference that can be made to other operating wireless systems.

A way to reduce the spectral lines of UWB signal is a random delay to the pulse. This can be seen from the PPM modulation which delays the pulse according to the random data. In Fig. 6.3 the PPM signal and its spectrum are shown. Comparing with Fig. 6.2, it is seen that the comb lines and peak powers can be reduced when PPM modulation is used.

Combining PAM and PPM can also be considered for modulating the UWB pulses, where the data is modulated in the amplitude as well as the delay of the UWB pulses. With this scheme higher data rates can be achieved.

6.1.2 Comparison of Data Modulation Schemes

According to [1], no serious attempt has been made to use either PAM or OOK for UWB, because in general, an amplitude-modulated signal which has smaller amplitude is more susceptible to noise interference than its larger amplitude counterpart. Furthermore, more power is required to transmit the higher amplitude pulse. One of the advantages of BPM modulation to the binary PPM modulation is the 3 dB gain in power efficiency. BPM has advantages like less susceptibility to distortion because the difference between two pulse levels is twice the pulse amplitude. Meanwhile, change of polarity can remove the PSD spectral lines as changing of polarity of pulses produces a zero mean [2]. However, BPM only supports binary communication. PPM modulation is efficient in minimizing the spectral peaks caused by pulse repetition in time. This can be further enhanced by using Time-Hopping technique, or by having more levels in the modulation. Pulse-position modulated UWB signals are less sensitive to channel noise compared to PAM signals. However, the main disadvantage of PPM is its sensitivity to timing synchronization.

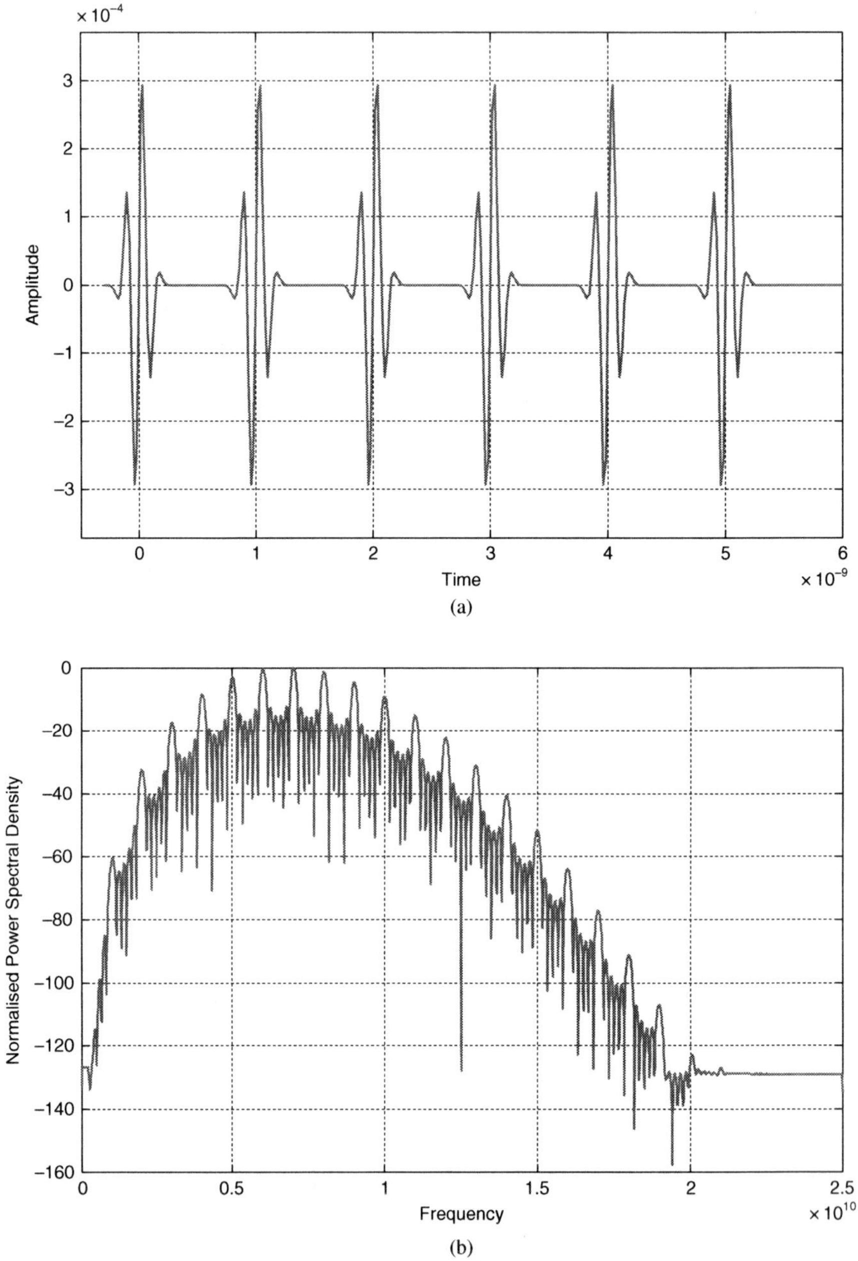

Fig. 6.2 A UWB pulse train in (**a**) time and (**b**) frequency domains

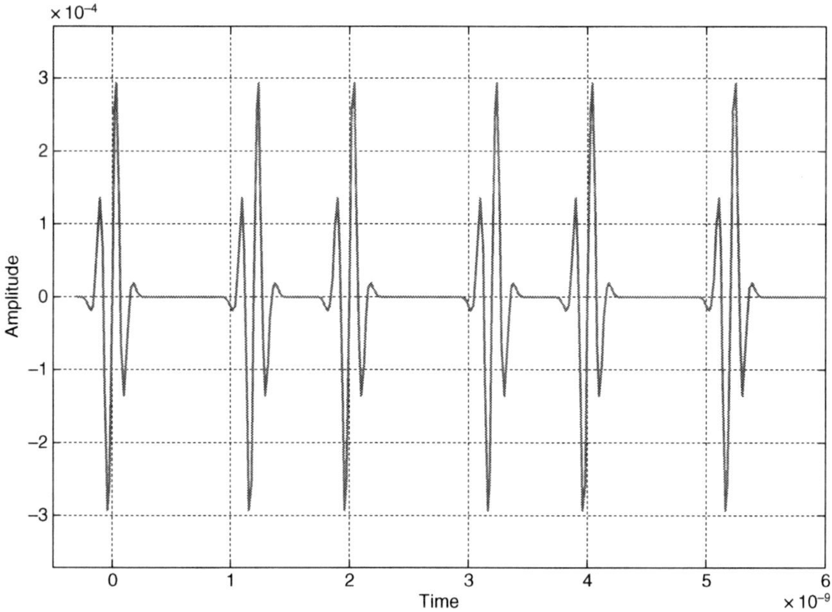

(a)

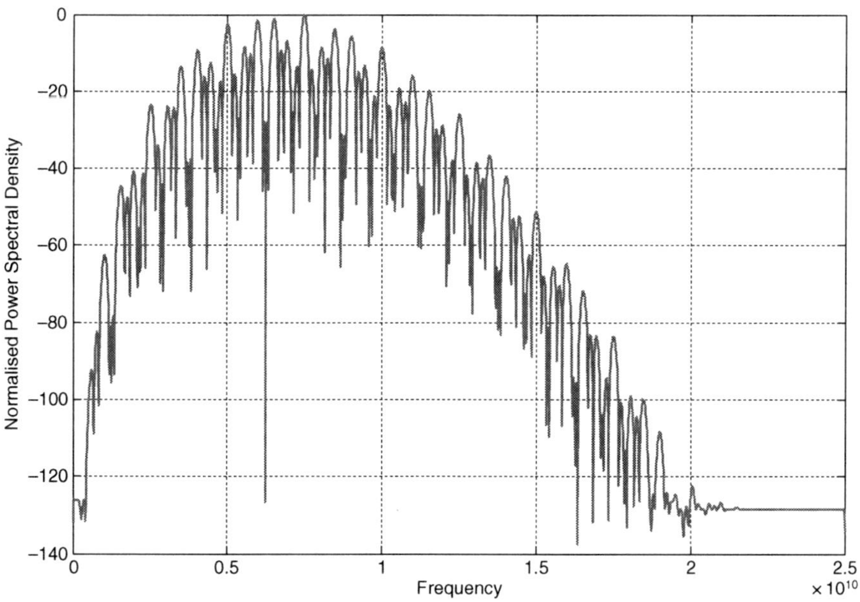

(b)

Fig. 6.3 (**a**) Pulse position modulation (PPM) and its (**b**) spectrum

6.1.3 UWB Multiple Access Modulation

Different modulation schemes can be used in the UWB communication system. The most attractive one is the Time Hopping (TH). A typical transmitted TH UWB signal can be given as [3]–[4]:

$$s_{tr}^{(k)}(t) = \sum_{i=-\infty}^{+\infty} p\left(t - iT_f - c_i^{(k)}T_c - \delta_{shift}D_{\lfloor i/N_s \rfloor}^{(k)}\right) \tag{6.5}$$

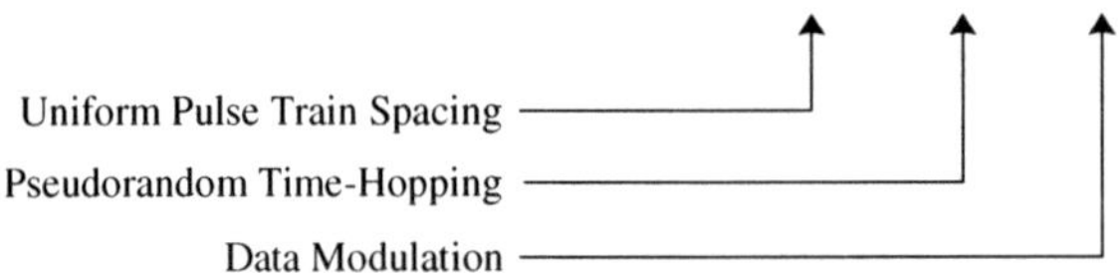

where $S_{tr}^{(k)}$ is the transmitted signal of user k, p is the generated pulse waveform, T_f is the pulse repetition interval, $c_j^{(k)}$ is a pseudorandom code, T_c is the chip time, $D \in \{0, 1\}$is the data sequence, N_s is number of pulses representing one data bit and δ_{shift} is the extra time shift (see also Fig. 6.1-c).

6.1.4 Uniform Pulse Train Spacing

The time between two-successive pulses is T_f and is referred as to the frame time. T_f can vary from a hundred to a thousand times the UWB pulse width. If we consider that a frame consists of M compartments, then it is better to give each compartment to each user as depicted in Fig. 6.4. However, if two or more users will occupy the same compartment (user “1” and user “k-1”), a catastrophic collision may occur because the pulses are uniformly spaced.

In addition to this, as the pulses are sent at regular intervals, the undesired peak power spectral lines are observed in the frequency domain at the pulse repetition rate as shown in Fig. 6.2.

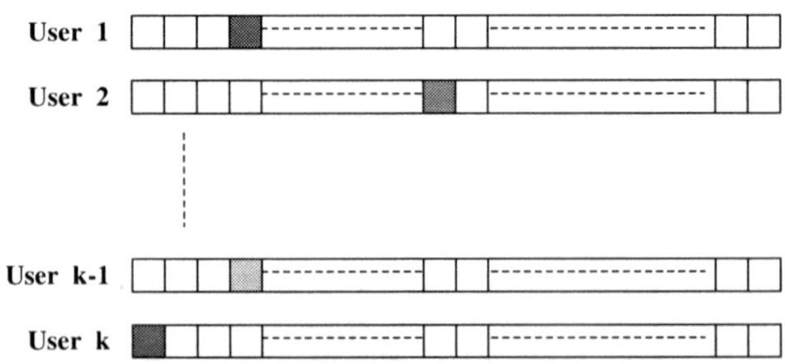

Fig. 6.4 Occurrence of the collision between pulses

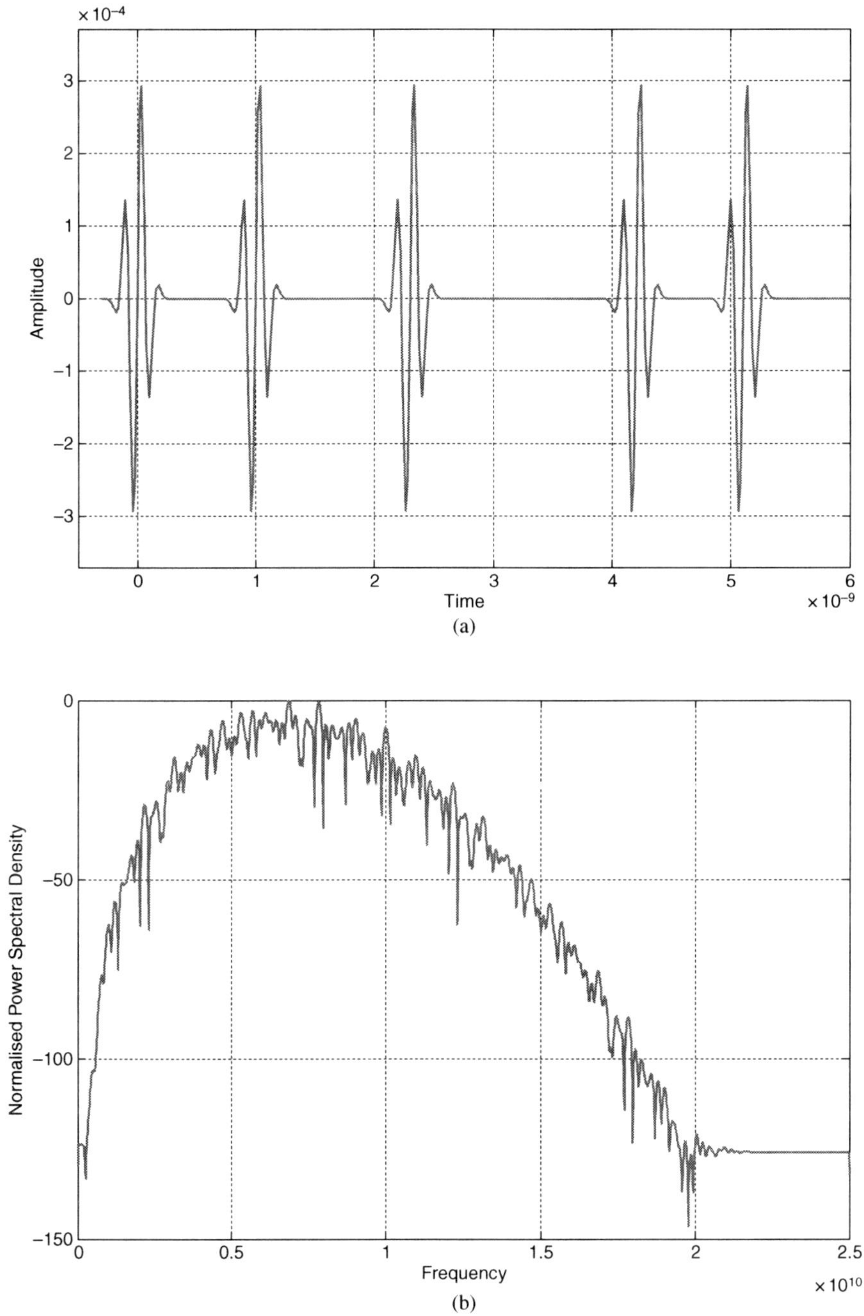

Fig. 6.5 The impact of pseudo-random time modulation (**a**) on time pulses, (**b**) on the energy distribution in the frequency domain

6.1.5 Pseudorandom Time Hopping

The solution to avoid catastrophic collisions of pulses is to assign each user its own code. This code is referred to as a periodic pseudorandom time hopping code $\{c_j^{(k)}\}$ with period N_p as expressed in (6.5). According to the code, each pulse will be randomly shifted with $c_j^{(k)} T_c$. As illustrated in Fig. 6.5 this randomness helps smoothing the spectrum (i.e., less peak power) and consequently less interference may occur to other communication systems.

6.1.6 Direct Sequence UWB (DS-UWB)

The multiple access capability can be provided by Direct Sequence (DS) spreading of the signal using a spreading signal. For the user k the spreading signal is:

$$c_k(t) = \sum_{i=-\infty}^{\infty} c_{k,i}\ P_{T_c}(t - iT_c) \tag{6.6}$$

The sequence $c_{k,i}$ is a periodic sequence of elements $\{+1,-1\}$ and $P_{T_c}(t)$ is a narrow time limited pulse with duration of T_c. Signal of kth user is expressed as:

$$b_k(t) = \sum_{i=-\infty}^{\infty} b_{k,i}\ P_T(t - iT) \tag{6.7}$$

where $b_{k,i} \in \{-1,1\}$ is the binary data of the kth user at ith interval T. The $P_T(t)$ has the shape like $P_{T_c}(t)$, but it has a duration of $T >> T_c$. The ratio of $T/T_c = M$ is called processing gain. Unlike the DS Code division multiple access (DS-CDMA), the processing gain of the DS-UWB systems is not usually large due to the high data rates of UWB transmission or (or small T). The kth user DS-UWB signal is

$$y_k(t) = c_k(t)\,.\,b_k(t) \tag{6.8}$$

$c_{k,i}$ in (6.6) is called signature sequence. It is a periodic sequence with the period of M , i.e., $c_{k,i+M} = c_{k,i}$. According to (6.8) in the DS-UWB system spreading of the data signal is carried out by direct multiplication of the data and spreading sequence. In the frequency domain the spectrum of the DS-UWB signal is the convolution of the spectrum of data and the spectrum of the spreading signal.

6.2 BER of Modulation Schemes

For M-ary PAM modulation, the symbol error probability (P_s) as function of E_b/N_0, where E_b is the average energy per bit and N_0 is the power spectral density height of additive white Gaussian noise (AWGN), can be obtained from Equation (6.9) [5] and can be seen from Fig. 6.6.

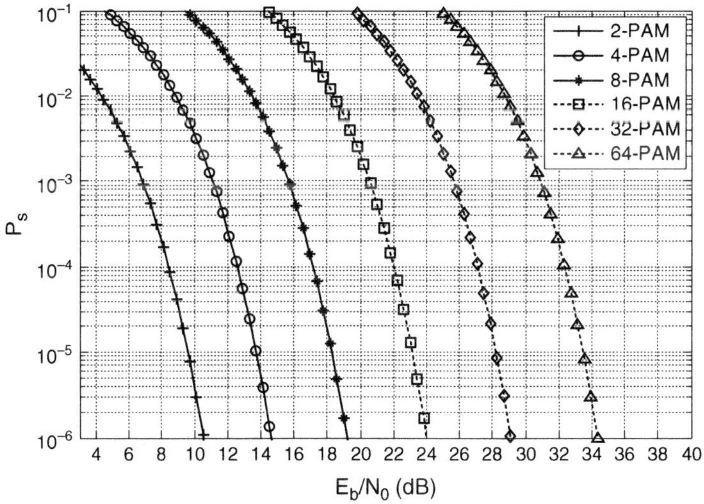

Fig. 6.6 Probability of a symbol error for M-ary PAM modulation as function of E_b/N_0 for AWGN channel

$$P_s = 2\left(1 - \frac{1}{M}\right) Q\left(\sqrt{\frac{E_b}{N_0}\frac{6\log_2 M}{M^2 - 1}}\right) \tag{6.9}$$

Assuming Gray mapping, the probability of bit error (P_{bit}) can be calculated as:

$$P_{bit} = \frac{1}{\log_2 M} P_s \tag{6.10}$$

and using (6.9)

$$P_{bit} = \frac{2}{\log_2 M}\left(1 - \frac{1}{M}\right) Q\left(\sqrt{\frac{E_b}{N_0}\frac{6\log_2 M}{M^2 - 1}}\right) \tag{6.11}$$

which has been shown in Fig. 6.7.

For M-ary PPM modulation, the symbol error probability P_s can be derived using an upper bound [5]–[7], as expressed by Equations (6.12), (6.13) and illustrated by Fig. 6.8.

The symbol error probability of M-ary PPM for $\frac{E_b}{N_0} > 4\ln 2$ is given by:

$$P_s < e^{\frac{-\ln M}{2}\left(\frac{E_b}{N_0} - 2\log_e(2)\right)} \tag{6.12}$$

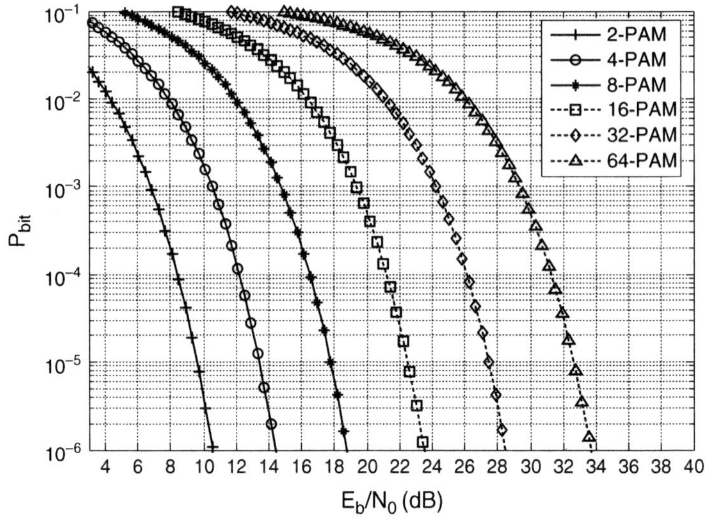

Fig. 6.7 Probability of a bit error for M-ary PAM modulation as function of E_b/N_0 for AWGN channel

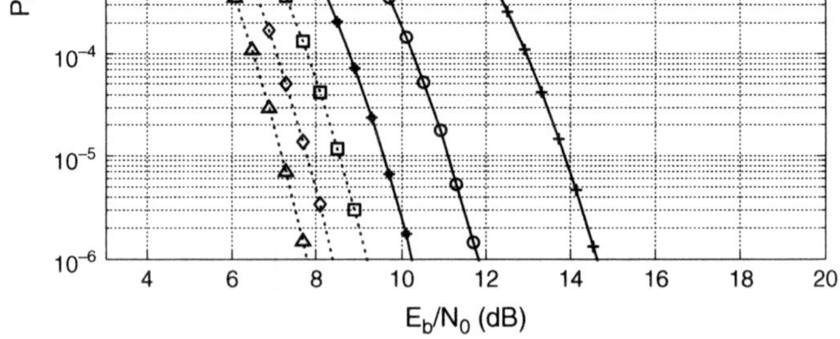

Fig. 6.8 Bounds of symbol error probability for M-ary PPM as function of E_b/N_0 for AWGN channel

and for $\frac{E_b}{N_0} < 4\ln 2$ and $\frac{E_b}{N_0} > \ln 2$ is given by:

$$P_s < 2e^{-\log_2 M\left(\sqrt{\frac{E_b}{N_0}} - \sqrt{\ln(2)}\right)^2} \tag{6.13}$$

The bounds of bit error probability P_{bit} of M-ary PPM modulation is obtained using

$$P_{bit} = \frac{2^{k-1}}{2^k - 1} P_s \tag{6.14}$$

where $k = \log_2 M$. For $\frac{E_b}{N_0} > 4\ln 2$ it is given by:

$$P_{bit} < \frac{2^{k-1}}{2^k - 1} e^{\frac{-\log_2 M}{2}\left(\frac{E_b}{N_0} - 2\log_e(2)\right)} \tag{6.15}$$

and for $\frac{E_b}{N_0} < 4\ln 2$ and $\frac{E_b}{N_0} > \ln 2$ is given by equation (6.16) and presented in Fig. 6.9.

$$P_{bit} < \frac{2^{k-1}}{2^k - 1} \cdot 2e^{-\log_2 M\left(\sqrt{\frac{E_b}{N_0}} - \sqrt{\log_e(2)}^2\right)} \tag{6.16}$$

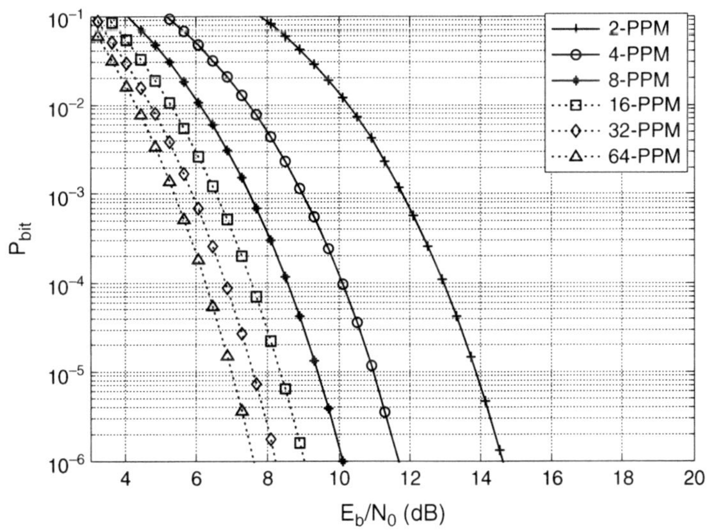

Fig. 6.9 Bounds of bit error probability for M-ary PPM as function of E_b/N_0 for AWGN channel

6.3 Rake Receiver

Rake receiver is the efficient receiver for the DS-UWB system. It takes advantage of the multipath effect of the wireless channels. As explained in Section 4.1, consider the impulse response of the kth user multipath channel as

$$h(t) = \sum_{n=1}^{N} a_n e^{j\theta_n} \delta(t - \tau_n) \tag{6.17}$$

If the bandwidth of the DS-UWB signal is denoted by $BW_{\text{DS-UWB}}$, which is significantly larger than the coherence bandwidth of multipath channel (Bc), the number of resolvable independent paths becomes:

$$N = \left\lfloor \frac{B_{DS-UWB}}{B_c} \right\rfloor + 1 \tag{6.18}$$

where $\lfloor \cdot \rfloor$ denotes the integer part of $(\cdot)$. In the Rake receiver the N resolvable paths are used in the detection of data. Because different paths are less likely to be in deep fading, the multipath diversity is well exploited and accordingly Rake receiver provides a better bit error rate performance. Rake receiver is the optimal receiver for the multipath channel. Its name comes from the garden rake having fingers to constitute the resolvable paths. Block diagram of the Rake receiver for user k is shown in Fig. 6.10. The received signal $y(t)$ is multiplied by the estimated channel coefficients in each Rake branch tuned to each resolvable path. For the optimum performance of Rake receiver the channel coefficient estimates should be the conjugate of the actual coefficient

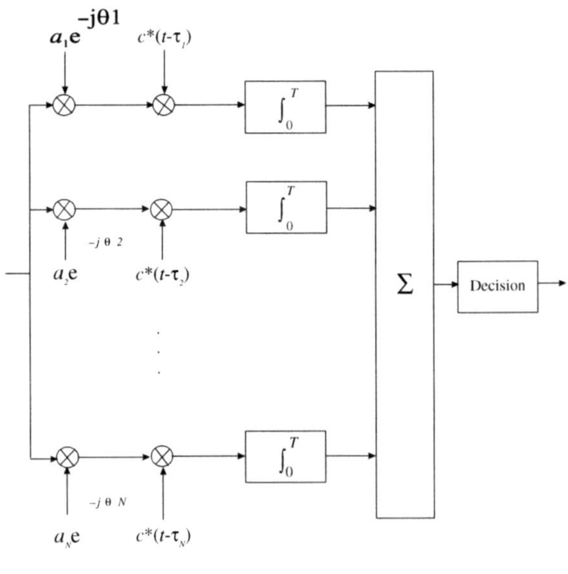

Fig. 6.10 Block diagram of the Rake receiver of kth user

of the appropriate paths. Deleting the index of kth user for the simplicity of derivations, the resulting signal in each Rake branch is then multiplied by the conjugate of the code sequence. After de-spreading by the codes, the outputs of correlators are combined to detect the data bit:

$$\hat{b} = \sum_{i=1}^{N}\left(\int_{0}^{T} y(t)a_i e^{-j\theta_i} c^*(t-\tau_i)dt + \int_{0}^{T} n(t)a_i e^{-j\theta_i} c^*(t-\tau_i)dt\right) \tag{6.19}$$

where in (6.19) T is the symbol duration, $c(t)$ is the spreading code, and $n(t)$ is AWGN. Using (6.7) and (6.8) the Eq. (6.19) can be written as:

$$\begin{aligned}\hat{b} = &\sum_{m=1}^{N}\sum_{i=1}^{N}\left(\int_{0}^{T} a_m e^{j\theta_m} a_i e^{-j\theta_i} b(t-\tau_m)c(t-\tau_m)c^*(t-\tau_i)dt\right.\\ &\left.+\sum_{i=1}^{N}\int_{0}^{T} n(t)a_i e^{-j\theta_i} c^*(t-\tau_i)dt\right)\end{aligned} \tag{6.20}$$

or

$$\hat{b} = \sum_{m=1}^{N}\int_{0}^{T} a_m{}^2 b(t-\tau_m)dt + \sum_{i=1}^{N}\int_{0}^{T} n(t)a_i e^{-j\theta_i} c^*(t-\tau_i)dt \tag{6.21}$$

The first term in (6.21) is the useful information and the second term is due to noise. As the data bit $b(t) = \pm A$ in the interval $[0,T]$ Eq. (6.21) can be simplified as:

$$\hat{b} = \pm AT\sum_{m=1}^{N} a_m{}^2 + v \tag{6.22}$$

where v is the noise term. As seen from (6.21) the first term is proportional to the sum of the power of the channel coefficients, while the noise term is proportional to the vectorial sum of the multiplied noise by the complex channel coefficients. Usually the real part of the first term is larger than the second term. Accordingly, the Rake receiver enhances the detection of data in the multipath wireless channels. Rake reception of DS-UWB signal and its enhanced performance is due to proper use of path diversity in the receiver. Better performance can be achieved by increasing the number of diversity paths. But this also increases the complexity of the UWB receiver.

There are two drawbacks of RAKE receivers for the UWB systems:

i) Because of extremely large bandwidth of the UWB signal a lot of Rake fingers will be required to capture energy, which increases the complexity of UWB receiver,

ii) As each multipath undergoes a different channel, this causes distortion in the received pulse shape which makes the use of a single LOS path signal as a template suboptimal [7].

6.4 Transmit-Reference (T-R) Technique

As mentioned in Section 6.3 for a lot of resolvable paths, the UWB rake receiver becomes complex. That is the number of parameters to be estimated (the number of delays and amplitudes) will be large. Another important issue is the sampling (at the subpulse rate) which is required to perform channel estimation. As an example with typical pulse length of $T_p = 700$ ps, the sampling rate of around $10/T_p \cong 14$ GHz is needed!

T-R method is a technique to avoid the high sampling rate and computational complexity of the estimation of channel $h(t)$. The block diagram of T-R transmitter is shown in Fig. 6.11-a. In this diagram the T-R method for transmission of PAM UWB signal with amplitude $a \in \{\pm 1\}$ is illustrated. In the T-R method each information pulse is coupled with an unmodulated pulse (pilot). We assume that the wireless channel has a maximum excess delay of T_{max} and the UWB pulse $p(t)$ has a duration of T_p. We also assume that the delay D is larger than the total pulse duration of pulse and the maximum delay spread of the channel, i.e., $D > T_{max} + T_p$.

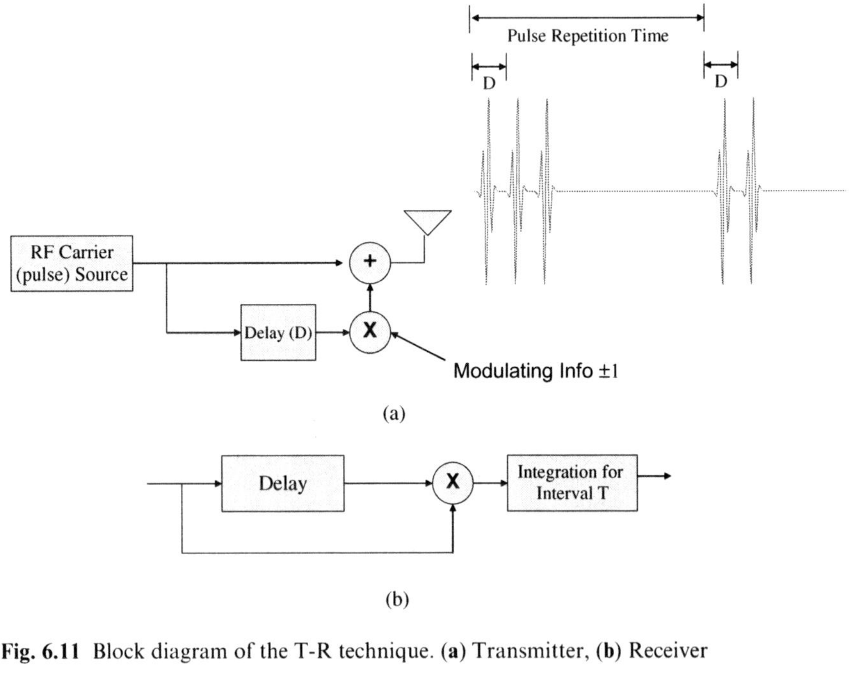

Fig. 6.11 Block diagram of the T-R technique. (**a**) Transmitter, (**b**) Receiver

Unlike conventional UWB transmission, in the T-R method a pair of UWB pulses separated by time interval D, which is known to both transmitter and receiver, is transmitted. The T-R transmit signal is written as:

$$v(t) = p(t) + ap(t - D) \tag{6.23}$$

Denoting the impulse response of the channel by $h(t)$, the received signal is expressed as:

$$r(t) = v(t)^*h(t) = [p(t) + ap(t - D)]^*h(t) = g(t) + ag(t - D) \tag{6.24}$$

where

$$g(t) = p(t)^*h(t) \tag{6.25}$$

In the receiver (see Fig. 6.11-b) detection of data is carried out by correlating the received signal with a replica of the received signal which is exactly delayed by the amount of delay in the transmitter (i.e., D):

$$\begin{aligned} R_{rr}(\tau) &= \int r(t)r(t - D)\mathrm{d}t = \int [g(t) + ag(t - D)][g(t - D) + ag(t - 2D)]\mathrm{d}t \\ &= R_{gg}(D) + aR_{gg}(2D) + aR_{gg}(0) + a^2R_{gg}(D) \\ &= 2R_{gg}(D) + aR_{gg}(2D) + aR_{gg}(0) \end{aligned} \tag{6.26}$$

With the assumption $D > T_{max} + T_p$ the received pilot and the information pulse do not overlap. The correlator receiver yields the symbol estimate as

$$\hat{a} = \mathrm{sgn}\left\{\int r(t)r(t - D)\mathrm{d}t\right\} \tag{6.27}$$

And accordingly in the absence of noise the data can be detected as

$$\hat{a} = \mathrm{sgn}\left\{a\int g^2(t)\mathrm{d}t\right\} = a \tag{6.28}$$

In the transmit-Reference method the same pulse is transmitted twice through the same channel. Both pulses experience the same channel distortion. The T-R receiver correlates the similarly distorted data pulse and reference pulse, which show high correlation. Therefore, in the T-R method there is no need for channel estimation algorithms [2]. Furthermore, in the T-R technique each reference pulse acts as a preamble for its data pulse, providing rapid synchronization. Another advantage of T-R method is when multipath components in the received signal contain significant energy. T-R receiver has the ability to capture that energy by correlating the received signal with its delayed version which is quite important in low-power UWB communications [2]. Despite the advantages of T-R technique it suffers from some drawbacks. Studies show that performance of T-R receivers is poor at low signal-to-noise ratios or in the presence of narrow band interference [8]. Another shortcoming of T-R technique is that in this method half of the transmitted waveform is used as pilot. Transmitting a reference pulse with each data pulse reduces the rate of data transmission.

The T-R method can be generalized by using more than one pilot pulse in the transmission. Major parameters of the generalized T-R method are the number of pilot pulse per burst, and the energy allocation among pilots and information pulses for maximum capacity. If N_f denotes the number of frames, T_f denotes the frame duration and N denotes the channel coherence time/T_f, the optimal number of pilots can be obtained as [9]

$$N_P = N - N_f(\lceil N/N_f \rceil - 1) \tag{6.29}$$

where $\lceil x \rceil$ is the ceiling of x or the closest integer $\geq x$. Another important factor is α which is the fraction of energy per burst assigned to pilot waveform. It is shown in [9] that for low SNR or small N the value of α is ½. While for large SNR or large N becomes:

$$\alpha = \frac{\sqrt{N_f}}{\sqrt{N - N_f} + \sqrt{N_f}} \tag{6.30}$$

6.5 UWB Range- Data Rate Performance

In this section we study the trade-off between range and data rate of the impulse radio UWB system. For the analysis of the power of received signal we use the free-space propagation channel model given by (4.27) which is repeated here:

$$PL(d) = \frac{(4\pi)^2 d^2 f_c^2}{G_r G_t c^2} \tag{6.31}$$

where PL is the path loss, d is the distance between UWB transmitter and receiver, G_t and G_r are respectively, the transmit and receive antenna gains (considered to be 0dBi), f_c is the carrier frequency, and c is the speed of light.

The total transmitted power P_t which complies with the FCC limits can be obtained by integrating the power spectral density of the UWB pulse in the frequency region of UWB:

$$P_t = \int_{-\infty}^{+\infty} A_{\max} |PSD_n(f)| df \tag{6.32}$$

where $A_{\max} = -41.3\,dBm/MHz$ and PSD_n is the normalized power spectral density of the nth derivative of Gaussian pulse. Assuming free space path loss, the received power at a distance d is obtained as:

$$P_r(d) = \frac{A_{\max} G_r G_t c^2}{(4\pi)^2 d^2} \int_{f_L}^{f_H} \frac{|PSD_n(f)|}{f^2} df \tag{6.33}$$

where f_L and f_H are respectively the lowest and highest UWB frequencies.

By considering additive and white Gaussian noise in the receiver, the received power necessary at a distance d to achieve a given average SNR_r can be computed [10]

$$P_r(d) = SNR_r.P_N.LM \tag{6.34}$$

where LM is the link margin, $P_N = N_0 B = kT_0 F.B$ is the noise power, k is the Boltzmann's constant (1.38×10^{-23} J/K), T_0 is room temperature (300 K), F is the noise figure and B is the noise equivalent bandwidth of the receiver.

If the symbol rate is assumed to be equal to the pulse repetition rate, a single UWB pulse is transmitted for each data symbol, and the energy per information symbol equals the energy per pulse. Hence, the average SNR in the receiver considering an ideal channel is expressed as:

$$\mathrm{SNR}_r = \frac{E_s/T_s}{N_0 B} = \frac{E_b R_b}{N_0 B} \tag{6.35}$$

where E_s is the received symbol energy, R_s is the symbol rate, and for non-coded systems, are given by $E_s = E_b \log_2 M$ and $R_s = R_b/\log_2 M$, respectively. E_b is the bit energy and R_b is the bit rate. For a specified target BER, the required (E_b/N_0) for M-ary PAM and M-ary PPM modulations can be obtained from (6.9) and (6.12), (6.13), respectively.

The range-data rate performance of IR-UWB system can be obtained as:

$$d = \frac{c}{4\pi} \sqrt{\frac{A_{\max} G_r G_t}{(E_b/N_0).R_b.kT_0 F.LM} \int_{f_L}^{f_H} \frac{|PSD_n(f)|}{f^2} df} \tag{6.36}$$

Some of the range-data rate performance results for different BERs for the PAM and PPM modulations are shown in Figs. 6.12 and 6.13, respectively. For obtaining these results an indoor UWB pulse of 7th order derivative of Gaussian pulse with spread of $\sigma = 51\,ps$ has been considered. Meanwhile, parameters such as $F = 6$ dB, $LM = 5$ dB (See (6.34)) have been assumed.

As seen from Figs. 6.12 and 6.13, for a fixed BER higher data rates can be achieved for shorter ranges. For a BER $= 10^{-3}$, with binary PAM data rate of

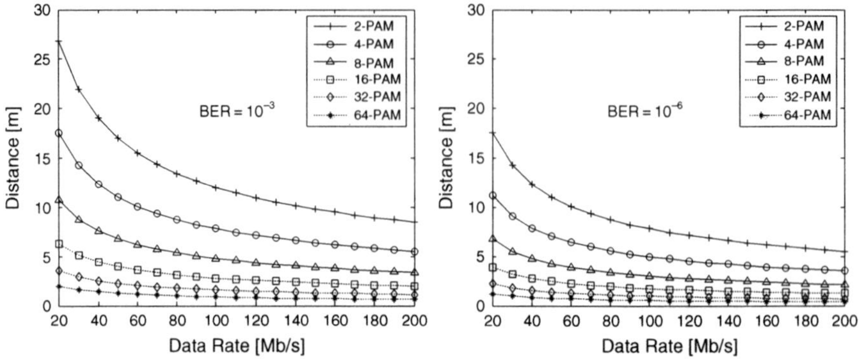

Fig. 6.12 Range-data rate performance of M-ary PAM UWB for different BERs

20 Mbps can be achieved at 27 m. While, for a data rate of 100 Mbps almost 12 m may be reached. For higher levels of PAM modulation, lower distances are achieved for the same data rate.

For lower BER criteria, and for the same modulation scheme, smaller distances can be reached. For example for the PAM modulation and BER of 10^{-6}, for a data rate of 20 Mbps, around 17 m may be reached, while for 100 Mbps around 8 m may be achieved.

As seen from Fig. 6.13 and unlike M-ary PAM case, the range-data rate performance of PPM system increases by increasing the modulation level. For the considered power limited system, longer ranges can be achieved by M-PPM when compared to PAM modulation. For instance, for BER = 10^{-6}, the range-data-rate performance of 4-PPM is lower than the performance of binary PAM. While, 8-PPM and higher PPM modulation levels can provide higher ranges comparing with the 2-PAM.

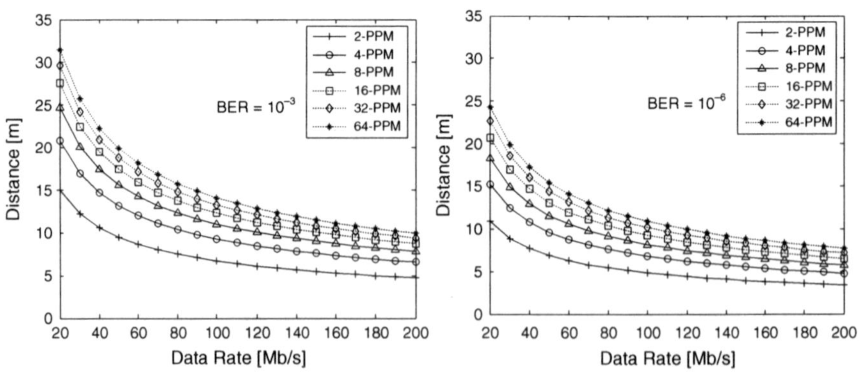

Fig. 6.13 Range-data rate performance of M-ary PPM UWB for different BERs

Results of Figs. 6.12 and 6.13 clearly show the trade off between the coverage and speed of data transmission of the UWB system. The presented results in these figures were for the indoor UWB. Complying with the FCC outdoor emission limit, the compromise between data rate and range can be further extended by using (6.32) and (6.36) showing the suitability of UWB transmission over longer ranges (several hundred of meters) at lower data rates (e.g., 100 kbps).

6.6 UWB Channel Capacity

UWB wireless systems use a very large bandwidth. The major advantage of large bandwidth is improved channel capacity. Channel capacity is defined as the maximum amount of data that can be transmitted per second over the channel. The large channel capacity of UWB systems is evident from the Shanon formula [2]:

$$C = B\log_2(1 + SNR) \tag{6.37}$$

where, C = Channel Capacity (bps), B = Bandwidth in Hz and SNR is the Signal-to-noise ratio. As seen from this expression by increasing B the channel capacity increases. Using several GHz of bandwidth for wireless UWB systems, a data rate of Gbps from this equation can be expected. However, due to FCC limit, high data rates can only be achieved for short ranges [2].

In this section we estimate the minimum power that is required at the UWB receiver to reach the Shannon limit [11]. Starting with (6.37) and noting the large bandwidth of UWB system (i.e., $B >> C$, by Taylor expansion of $\log(1+x) = x - x^2/2 + x^3/3! + \ldots$ and considering only the first term, (6.37) can be approximated as:

$$\frac{C}{B} \cong \frac{SNR}{\ln 2} \tag{6.38}$$

Signal-to-noise ratio SNR can be written as:

$$SNR = \frac{\text{Signal power}}{\text{Noise power}} = \frac{P}{kTB} \tag{6.39}$$

where P is the signal power, k = Boltzmann′s constant = 1.38×10^{-23} Joule/°K, and T is Temperature (Kelvin). Taking the logarithm of both sides of (6.39) and using (6.38) we have:

$$PdB = 10\log C + 10\log kT + 10\log(\ln 2) \tag{6.40}$$

SNR can be related to the energy per bit over spectral noise density height (E_b/N_0) using the rate of transmission C:

$$SNR = \frac{E_b}{N_0} \cdot \frac{C}{B} = \frac{E_b}{N_0}\rho \tag{6.41}$$

Where $\rho = C/B$. Now by substituting (6.41) in (6.37) we get:

$$\rho = \log_2\left(1 + \frac{E_b}{N_0}\rho\right) \tag{6.42}$$

By manipulating (6.42), the Shannon capacity limit can be written as

$$\frac{E_b}{N_0} = \frac{2^\rho - 1}{\rho} \tag{6.43}$$

As the bandwidth of UWB is very large the value of ρ is small. Accordingly, the Shanon limit can be achievable when minimum E_b/N_0 is ln2 as shown below.

$$\frac{E_b}{N_0} = \lim_{\rho \to 0} \frac{2^\rho - 1}{\rho} = \ln 2 \tag{6.44}$$

6.7 Summary

The various modulation schemes of impulse radio IR-UWB was presented in this chapter. This included UWB data and multiple access modulations, reception techniques such as Rake and Transmit-Reference receivers. Special emphasis was given to the performances of M-ary pulse amplitude modulation and Pulse position modulation.

Problems

Problem 6.1 Compare the BPM, PAM, PPM and T-R methods with each other and list their advantages and disadvantages for IR-UWB transmission.

Problem 6.2 A wireless channel has an impulse response as $h(t) = e^{-at}u(t)$, where $u(.)$ is the unit step function.

(a) The input signal to this channel is $x(t) = \cos 2\pi f_0 t$. Calculate the optimum value of a which maximizes the signal-to-noise ratio (SNR) out of the channel. Then compute this maximum SNR.
(b) Instead of the $\cos 2\pi f_0 t$ a UWB signal of narrow time pulses having a duration of τ and pulse repletion interval of $T_0 = 1/f_0$ is transmitted through this channel (see Fig. P.6.2). For $\tau/T_0 = 0.1$ and for the value of a found in part (a), calculate the output SNR and compare it with the SNR_{max} of part (a).

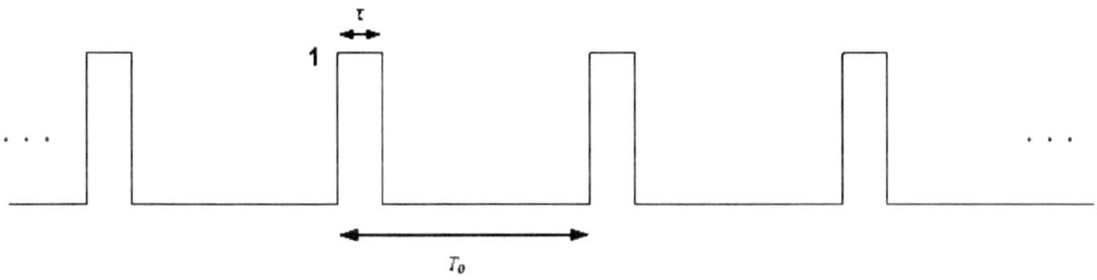

Fig. P.6.2 The UWB signal transmitted through the channel

Problem 6.3 In the T-R method the delay D used in the transmitter is the same as the delay used in the receiver. If delay used in the receiver is $\tau \neq D$, calculate the output of the correlator of the receiver. Consider a sinc function with bandwidth B centered at f_0 as the autocorrelation of $g(t)$ (see (6.25))

$$R_{gg}(\tau) = \text{sinc}(B\tau)\cos 2\pi f_0 \tau$$

Show that there is a leakage of energy from a carrier modulated at one delay to pulse pair correlator at another delay [12] due to sidelobes of the autocorrelation of transmitted signal. This leakage can be minimized by increasing bandwidth B. Show that the degree of mismatch delay in transmit and receive sides is tolerated can be controlled by f_0. Conclude that there is a direct relationship between a requirement for highly accurate delays and a higher center frequency. These points clearly guide us to UWB signals having very large bands and low center frequencies.

Problem 6.4 In this problem we study the effect of the narrowband interference on the performance of the T-R system [13]. Assume that the narrowband interference (NBI) can be expressed as

$$i(t) = \sqrt{2I}\cos(\omega_i t + \theta_i)$$

where I is the average power of NBI, at the frequency $f_i = \omega_i/2\pi$, and θ_i is the Uniform[0,2π) phase. Consider the impulse response of the channel as $h(t)$. Evaluate the decision variable at the output of the correlator in the receiver and show that the NBI introduces a bias term to this variable. How this bias term can be suppressed?

Problem 6.5 In Section 6.4 the range-data rate performance of IR-UWB was evaluated using the free space path loss. For the following dual slope path loss model

$$\text{path loss exponent} = \begin{cases} 1.8 & d < D_{bp} \\ 4 & d \geq D_{bp} \end{cases}$$

where D_{bp} is the breakpoint and assumed to be 50 m. Obtain the range-data rate performance of binary PAM UWB system for BER = 10^{-6} and compare it with the shown results of section 6.4. Repeat problem for the binary PPM and compare it with the binary PAM case.

Problem 6.6 In this problem we want to study the effect of the pulse shape on the range-data rate performance of the UWB system. Consider the UWB waveform $p(t)$ as the nth order derivative of Gaussian shape.

$$p(t) = \frac{d^n}{dt^n}\left(\frac{A}{\sigma\sqrt{2\pi}}\exp(-t^2/2\sigma^2)\right)$$

A is the amplitude of the pulse and the he total duration of the pulse is considered to be T_p=1 nsec. Complying with the indoor FCC mask, for n=1,3,5 and 7, obtain the proper value of A and the spread of the pulse σ. Then evaluate the range-rate performance of the system with different n. Draw your conclusion.

Problem 6.7 In the UWB systems usually transmitter and receiver antennas operate as differentiator. Consider the UWB transmitted pulse as $p_{TX}(t)$ having the Gaussian shape [14]:

$$p_{TX}(t) = \exp\{-2\pi(t/\tau_m)^2\}$$

Where the parameter τ_m is the width of the pulse and is assumed to be 75 psec.

(a) Obtain the received UWB pulse, $p(t)$, and plot it. Assume a duration of 2 ns for the received pulse (T_p = 2 ns).
(b) Determine the Energy of the received pulse as

$$E = \int_{-\infty}^{\infty} p^2(t)dt$$

(c) Assume that binary information signals are defined using pulse position modulation in such a way that for the bit "1" and in the interval of $0 \le t \le T$ = signal duration, $p_{TX}(t)$ is transmitted and for bit "–1" $p_{TX}(t\text{-}\tau)$ is sent. Similarly the received signals for bits "1" and "–1" are respectively, $p(t)$ and $p(t\text{-}\tau)$, $0 \le t \le T$ and $T > T_p + \tau$. Evaluate the squared distance between received signals as:

$$d^2(\tau) = \frac{1}{2E}\int_{-\infty}^{\infty} [p(t) - p(t-\tau)]^2 dt$$

Calculate $d^2(\tau)$ and sketch it for $0 < \tau < 2$ ns.

The probability of error can be obtained by

$$P_{error}(\tau) = Q\left(\sqrt{\frac{1}{2}d^2(\tau).SNR}\right)$$

where $Q(.)$ is

$$Q(x) = \frac{1}{\sqrt{2\pi}}\int_{x}^{-\infty} e^{-u^2/2}du$$

Using the plot of $d^2(\tau)$ find the value of τ that minimizes the error probability. Then determine the minimum bit error probability for SNR = 10 dB.

References

1. M. Ghavami, L.B. Michael and R. Kohno, Ultra Wideband Signals and Systems in Communication Engineering, Wiley, 2004.
2. F. Nekoogar, Ultra-Wideband Communications, Prentice-Hall, NJ. 2006.
3. M.Z. Win and R.A. Scholtz, "Impulse radio: How it works", IEEE Communications Letters, vol. 2, no. 2, February 1998, pp. 36–38.
4. M.Z. Win and R.A. Scholtz, "Ultra-wide bandwidth time-hopping spread-spectrum impulse radio for wireless multiple-access communications", IEEE Transactions on Communications, vol. 48, no. 4, April 2000, pp. 679–689.
5. J. Proakis, Digital Communications, 2nd Edition, McGraw Hill, NY, 1989.
6. M. Benedetto, G. Giancola, Understanding Ultra Wide Band Radio Fundamentals, Prentice Hall, 2004.
7. J.H. Reed, An Introduction to Ultra Wideband Communication Systems, Prentice Hall, NJ., 2006.
8. F. Dowla, F. Nekoogar and A. Spiridon, "Interference mitigation in Transmit-Reference Ultra-Wideband receivers," IEEE International Symposium on Antenna and Propagation, 2004, pp. 1307–1310.
9. L. Yang and G.B. Giannakis, "Ultra wideband communications, an idea whose time has come," IEEE Signal Processing Magazine, November 2004, pp. 27–54.
10. H. Sheng et al., "On the spectral and power requirements for Ultra-Wideband transmission," 2003 IEEE International Conference on Communications (ICC '03), vol. 1, 11–15 May 2003, pp. 738–742.
11. F. Dowla, Handbook of RF and Wireless Technologies, Elsevier, Amsterdam, 2004. Chapter 16.
12. R. Hoctor and H. Tomlinson, "Delay-hopped transmitted-reference RF communications," IEEE Conference on UWB Systems and Technologies, 2002, pp. 265–2269.
13. M. Pausini and G. Janssen, "Narrowband interference suppression in Transmit-Reference UWB receivers using sub-band notch filters," EUSIPCO 2006, Italy.
14. F.R. Mireles, "Signal detection for Ultra-Wide-Band communications in dense multipath," IEEE Transaction on Vehicular Technology, vol. 51, no. 6, November 2002, pp. 1517–1521.

Chapter 7
UWB Technologies

Contents

The Ultra WideBand (UWB) is a hot topic in wireless communications. In the near future this technology may see increased use for high-speed short range wireless communications, ranging and ad hoc networking. There are two competing technologies for the UWB wireless communications, namely: Impulse Radio (IR) and Multi-band OFDM (MB-OFDM). IR technique is based on the transmission of very short pulses with relatively low energy. The MB-OFDM approach divides the UWB frequency spectrum to multiple non-overlapping bands and for each band transmission is OFDM. Several proposals based on these two technologies have been submitted to the IEEE 802.15.3a. Both technologies are valid and credible. In this chapter these technologies will be discussed in more detail and major characteristics of each technique will be contemplated. Further, the two technologies will be compared from different aspects such as channel, interference, performance and complexity point of view. The structure of this chapter is as follows. In Section 7.1 the Impulse Radio technology is explained and its advantages and drawbacks are discussed. In Section 7.2 the pulsed multiband approach is studied. The UWB system based on combining OFDM with multibanding is explained in Section 7.3, and finally the IR and multiband OFDM UWB technologies are compared in Section 7.4

7.1 Impulse Radio

In impulse radio UWB pulses of very short duration (typically in the order of sub-nanosecond) are transmitted. Because of very narrow pulses the spectrum of the signal reaches several GHz of bandwidth. The impulse radio UWB is a carrier-less transmission. This technology has a low transmit power and

H. Nikookar, R. Prasad, *Introduction to Ultra Wideband for Wireless Communications*, Signals and Communication Technology,
DOI 10.1007/978-1-4020-6633-7_7,

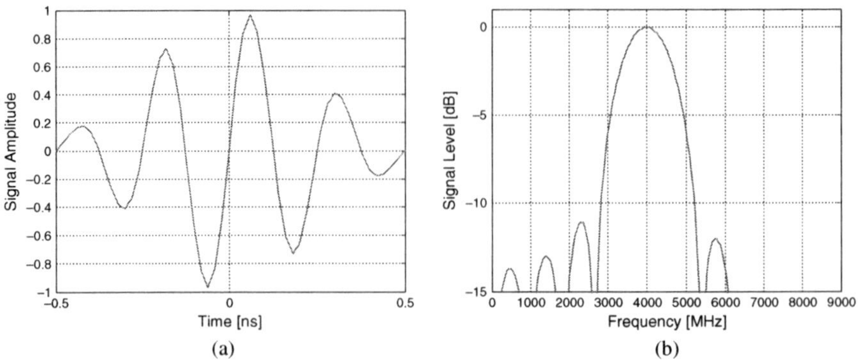

Fig. 7.1 UWB IR (**a**) and its spectrum (**b**) in the 3.1–5 GHz range

because of narrowness of the transmitted pulses has a fine time resolution. The implementation of this technique is very simple as no mixer is required which means low cost transmitters and receivers. In Fig. 7.1 the pulse shape in time domain and its spectrum is shown (3.1–5 GHz). Figure 7.2 illustrates another sample pulse shape, which occupies the 6–10 GHz band. The lowest cost systems are envisaged to use the lower frequency bands. The two bands can be used independently, or together, to provide a range of options for system deployment [1].

Direct Sequence Ultra Wideband (DS-UWB) and Time Hopping Ultra Wideband (TH-UWB) are two variants of the IR technique. These IR techniques were described in Section 6.1. DS-UWB and TH-UWB are different multiple access techniques that spread signals over a very wide bandwidth. Because of spreading signals over a very large bandwidth, the IR technique can combat interference from other users or sources. It should be mentioned that Direct Sequence Spread Spectrum (DSSS) and Time Hopping Spread Spectrum (THSS) may be

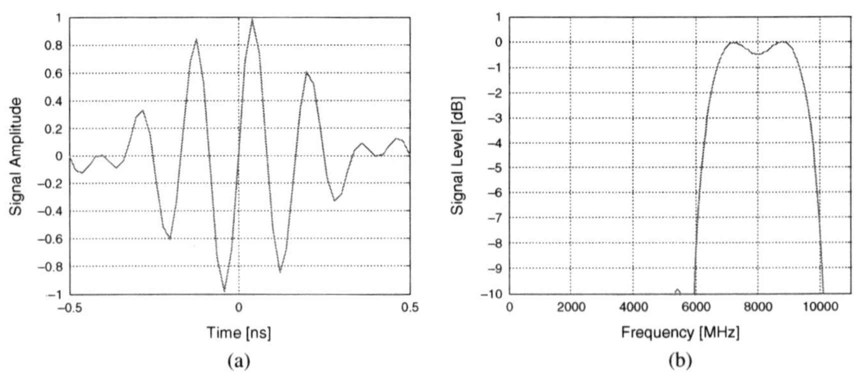

Fig. 7.2 UWB IR (**a**) and its spectrum (**b**) in the 6–10 GHz range

Table 7.1 IR-UWB system parameters for the 3.1–5.1 GHz, [1]

Information data rate	112 Mbps
Symbol rate	42.75 Msym/sec
Coding rate	0.44
Code length	32 chips/sec
Channel chip rate	1.368 Gcps
Modulation/constellation	64-Biorthogonal Keying

considered similar to DS-UWB and TH-UWB, respectively. There are, however, differences between the spread spectrum and IR-UWB systems. Both systems take advantage of the expanded bandwidth, while different methods are used to obtain such large bandwidth. In the conventional spread-spectrum techniques, the signals are continuous-wave sinusoids that are modulated with a fixed carrier frequency, while in the IR-UWB (i.e., DS-UWB and TH-UWB), signals are basically baseband and the narrow UWB pulses are directly generated having an extremely wide bandwidth. Another difference is the bandwidth. For the UWB signals the bandwidth has to be higher than 500 MHz, while for the spread spectrum techniques bandwidths are much smaller (usually in the order of several MHz).

One possible set of IR-UWB system parameters for the 3.1–5.1 GHz is shown in Table 7.1.

7.1.1 Complexity

UWB based on IR technology is a low-complexity radio technology as it is carrier-less and needs no mixer.

7.1.2 Power Consumption

If the coverage area is not large the transmit power of the IR can be kept low. Other important issue in the IR technology is the impulsive nature of transmission. This allows low transmit power when compared to the continuous transmission systems. It should also be noted that because of the impulsive nature of the IR UWB transmission, the multiuser interference may substantially differ from the continuous transmission. This again allows low transmit power when compared to continuous transmission systems.

7.1.3 Security

For the wireless applications, if needed, the security mechanism should be implemented in the physical layer of the UWB-IR.

7.1.4 IR Industry Standard Groups

In March 2005 the IEEE 802.15 has adopted as a baseline proposal for a draft standard UWB technology using impulse radio and the opportunity of supporting different kind of receivers. This has been adopted after successful merge of 24 proposals, 22 of each were favoring impulse radio type of UWB technology. This baseline proposal has been adopted by 100% of the participants. Among the proposers are STM, Freescale, Hitachi, Fujitsu, Samsung, Time Domain, CEA-Leti, NICT and France Telecom.

7.1.5 Other Features of IR Technology

Another important feature of the UWB IR is its fine time resolution, which provides distancing and location capabilities to wireless networks. Accurate distance measuring of the UWB-IR has been demonstrated in the recent years [1], [2].

The main disadvantages of the IR-UWB technology are the following:

- Challenge in building RF circuits with extremely large bandwidth,
- Achieving high-speed A/D and D/A is difficult to process the extremely large bandwidth,
- Digital complexity to capture multipath energy in dense multipath environments.

7.2 Pulsed Multiband

The idea behind multi-band approach is the use of multiple non-overlapping frequency bands to efficiently exploit UWB spectrum by transmitting single carrier or multicarrier signals. Ideally, signals do not interfere among themselves, because they operate at different frequencies.

There are several advantages of using multi-band UWB techniques, such as the ability to efficiently utilize the entire 7.5 GHz of spectrum made available to the UWB systems, with the use of appropriate multi-bandwidths. Separate bands along the available spectrum may be treated independently and benefits in coexistence with other interfering systems operating in the same band can be achieved. In the Pulsed Multi-band technique the large bandwidth is divided into several smaller subbands (e.g., 500 MHz band) and single carrier modulation is used for transmission in each subband. By interleaving the symbols across the subbands, the UWB system can still maintain the same transmit power as if it were using the entire band [3]. Major drawbacks of pulsed multiband technique are:

- Difficulty in collecting significant multipath energy using a single RF chain,
- The amount of multipath energy is limited by the dwell time of each subband,
- With small number of RAKE fingers the system performance is sensitive to group delay variations introduced by front end components,
- Pulsed multi-band needs a very fast band switching timing requirements (<100 psec) at both transmitter and receiver.

7.3 Multiband OFDM

Multi-band Orthogonal Frequency Division Multiplexing (MB-OFDM) is another UWB technology which uses the OFDM method. Multi-Band OFDM combines the OFDM technique with the multi-band approach. The spectrum is divided into several sub-bands with a –10 dB bandwidth of at least 500 MHz. The information is then interleaved across sub-bands and then transmitted through multi-carrier (OFDM) technique.

One of the proposals for the physical layer standard of future high speed Wireless Personal Area Networks (WPANs) uses MB-OFDM technique and is presented in [4]. In this MB-OFDM WPANs proposal, the spectrum between 3.1 and 10.6 GHz is divided into 14 bands with 528 MHz bandwidth that may be added or dropped depending upon the interference from, or to, other systems. In Fig. 7.3 a possible band plan is presented, where only 13 bands are used to avoid interference between UWB and the existing IEEE 802.11a signals. The three lower bands are used for standard operation, which is mandatory, and the rest of the bands are allocated for optional use or future expansions.

The proposed physical layer for UWB [5] supports variable data rates up to 480 Mbps. The desired range is 10 m for 110 Mbps and 2 m for 480 Mbps. The power consumption is in the order of the hundreds of milliwatt, to enable wireless connectivity for battery-operated portable devices, and varies with the bit rate. The target bit error rate is equal to 10^{-5}.

Transceiver architectures for MB-OFDM systems are similar to the conventional wireless OFDM systems. The main differences for MB-OFDM are in the use of time-frequency codes to specify center frequencies for the transmission of

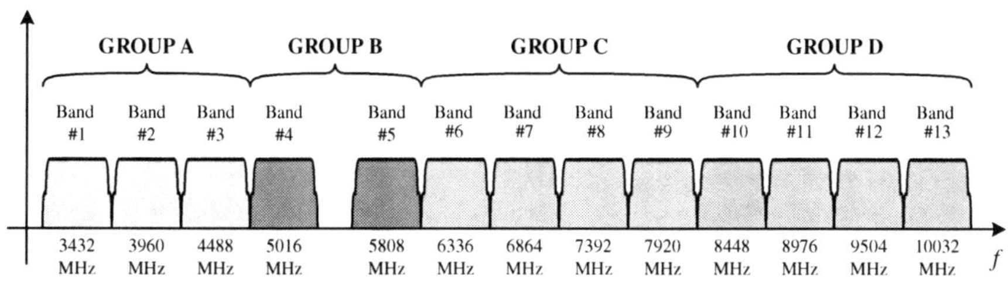

Fig. 7.3 Proposed MB-OFDM frequency band plan [5]

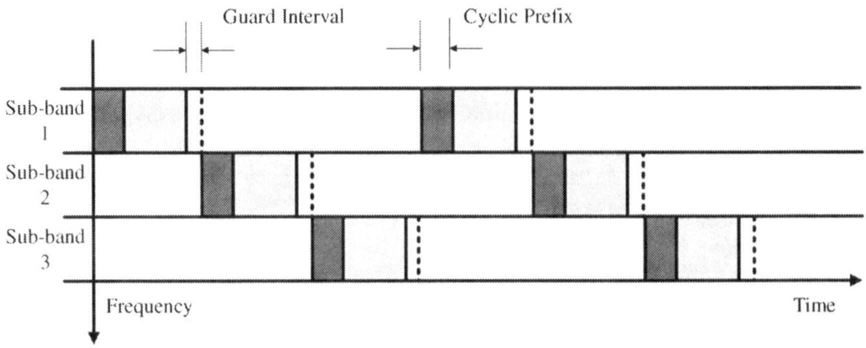

Fig. 7.4 Example of time–frequency coding for the MB-OFDM system's proposal [6]

each OFDM symbol [6]. Figure 7.4 illustrates this procedure for 3 of the 14 possible sub-bands. Considering transmission over time, the first OFDM symbol is transmitted on sub-band 1, the second OFDM symbol on sub-band 2, the third OFDM symbol on the sub-band 3 and this repetition continues over time. In practice, the frequency-time codes that specify the frequency repetition over time may be different to the one presented in Fig. 7.4. Furthermore, multiple access is specified by different time-frequency codes. The system parameters of a MB-OFDM system are shown in Table 7.2. The QPSK modulation is used for the OFDM tones and a 128-point FFT generates the OFDM signal. Additional 500 MHz wide channels provide additional system capacity.

As can be seen in the Fig. 7.4, a cyclic prefix (CP) is inserted before each OFDM symbol. The length of the CP determines the amount of multipath energy that can be captured. The CP length needs to be chosen to minimize the impact of Inter-Carrier Interference (ICI) and maximize the collected multipath energy, while keeping the overhead due to the CP small. Most conventional wireless OFDM-based systems use a CP to provide robustness against multipath. However, similar multipath robustness can be obtained by using a Zero-Padded Prefix (ZPP) instead of the CP [7]. The use of the CP increases

Table 7.2 OFDM system parameters for a UWB system [1]

Data rate	110 Mbps
Modulation Constellation	QPSK
FFT size	128
Information tones	50
Information length	242.4 ns
Guard interval	9.47 ns
Symbol interval	312.5 ns
Coding rate	11/32
Data tones	100
Spreading rate	2

robustness to multipath dispersion with a low-complexity receiver. Using ZPP the power back off at the receiver can be avoided. The power spectral density of OFDM signal using CP has ripples (resulting in transmit power back off of about 1.5 dB). While the ZPP provides no ripples in the power spectral density and accordingly no power back off is needed [6], [8]. As specified in Fig. 7.4, the guard interval ensures orthogonality among channels and that the switching between the different sub-bands can be accomplished within a few nanoseconds.

MB-OFDM signals include processing of the information over much smaller bandwidth compared to the 7.5 GHz made available to UWB systems. One of the remarkable advantages of of the MB-OFDM system is that the multi-band design of the OFDM system allows the technology to cope with local regulations by dynamically turning off some carriers to comply with local rules of operation on the allocated spectrum. By avoiding interferer's bands, it is also possible to simplify the design of the system.

MB-OFDM systems may have some problems related to the use of the OFDM technology. Two major problems are handling large peak-to-average power ratios (PAPR) and the synchronization issue. The summing of orthogonal frequencies in the Fast Fourier Transform (FFT) causes high PAPR of the OFDM signal. A peak in the signal power will occur when sub-carriers align themselves in phase. High PAPR is not suitable for the RF power amplifiers that have to work in the non-linear regions. The traditional method to mitigate the distortion caused by the PAPR is to use a linear high power amplifier, however this kind of solution does not fulfil the power requirements. Signal scrambling and signal distortion techniques have been proposed to mitigate the PAPR problems of OFDM. Some of signal scrambling algorithms give a very low PAPR, however, they are normally difficult to be realized in practice. Solutions like block coding [9], selected mapping [10], partial transmit sequences [11], envelope clipping [12] and weighting [13] are suggested to reduce the PAPR.

MB-OFDM technique is also very sensitive to the inter-modulation distortion that results from non-linearities in the RF amplifiers. Because the sub-carriers are equally spaced, the third-order intermodulation products will appear exactly on top of another carrier. These intermodulation products will contribute to a noise-like cloud surrounding each constellation point. Due to this fact, the lower-level modulations should be used. If not, these constellation clouds can contribute to an increase in bit errors for each carrier. The sub-carriers are perfectly orthogonal if transmitter and receiver use exactly the same frequencies. Any frequency offset results in ICI. Phase noise will also cause ICI in the MB-OFDM receiver. Sensitivity to phase noise and frequency offset are main disadvantages of OFDM systems in general and MB-OFDM in particular.

MB-OFDM technology promises to deliver data rates of about 110 Mbps at a distance of 10 m. For the UWB wireless sensor applications data rates are low, but the (hoping) coverage might be much larger than 10 m. MB-OFDM may require higher power levels when compared to the IR technology. MB-OFDM technique is robust to multipath which is present in the wireless channels. By using this technology and limiting the transmitting symbols to QPSK constellation, the resolution of the A/D and D/A and the internal precision of the digital baseband, especially the FFT, can be lowered. This technology has been implemented for the UWB communication based on CMOS technology [14].

7.3.1 Complexity

MB-OFDM basically is a low-complexity radio technology. OFDM is implemented by the FFT. By using the QPSK constellation the resolution of A/D and D/A can be reduced. Increasing the spacing between sub-carriers relaxes the phase noise requirement of OFDM technology and improves the synchronization errors.

7.3.2 Power Consumption

Battery life of the MB-OFDM devices working in wireless environments is a critical factor. MB-OFDM is capable of supporting a minimum of two hours continuous battery usage before a recharge is required [14]. The estimated power consumption for MB-OFDM system implemented in a 90 nm CMOS technology is reproduced in Table 7.3. It should be noted that in some applications such as UWB wireless sensor networks the data rates are not large. This will further reduce the required power levels of this radio technique.

7.3.3 Security

If needed, the security mechanism can be implemented at several stages of the protocol stack of the MB-OFDM technique as this technique can provide an embedded always "on" secure foundation.

7.3.4 Costs

Adoption of MB-OFDM technique for the wireless networks will depend on the ease of use of the technology at an affordable cost. In this regard the importance of low complexity CMOS solutions should not be ignored.

Table 7.3 Estimated power consumption of MB-OFDM system [4]

Data rate	Transmit power	Receive power	Deep sleep power
110 Mbs	93 mW	155 mW	15 μW
200 Mbps	93 mW	169 mW	15 μW

7.3.5 MB-OFDM Industry Standard Groups

Multiband OFDM alliance was established in 2003 which tries to promote the global standard for this technology of UWB wireless solutions [15]. Over 170 members participate in this alliance including the semiconductor, personal computing, consumer electronics and mobile devices companies around the world.

7.3.6 Other Features of MB-OFDM Technology

One of the features of MB-OFDM is its spectral flexibility. This feature is very important for the usage of this radio technology in the environments full of interferers. In the interference rich scenarios the MB-OFDM is capable of complying with the regulations and dynamically turning off certain carriers or channels in software. This flexibility is not provided by the IR technology.

The advantages of the MB-OFDM technique are summarized below [8]:

- Capturing multipath energy with a single RF chain
- Insensitivity to group delay variations
- Ability to deal with narrowband interference at receivers
- Simplified synthesizer architectures relaxing the band switching timing requirements

And the disadvantages are as follows:

- Transmitter is more complex because of IFFT
- High peak-to-average power ratios
- OFDM synchronization problems

7.4 Comparison of UWB Technologies

In this section the comparison of IR DS-UWB and MB-OFDM UWB techniques in terms of interference from, or to, other systems, robustness to multipath, performance, system's complexity and achievable range-data rate performance for the WPAN applications, is provided. It should be mentioned that the comparison here is by no means a comprehensive review of the capabilities of both technologies, but rather an attempt to appraise the possibilities and potentials of each technique. Both MB-OFDM and IR DS-UWB proposals are summarized in Table 7.4, [4] and [16].

Considering the MB-OFDM proposal, for data rates from 55 to 200 Mbps a time-domain spreading operation is performed with a spreading factor of 2. The time-domain spreading operation consists of transmitting the same information over two OFDM symbols. These two OFDM symbols are transmitted over different sub-bands to obtain frequency diversity [4].

Table 7.4 Specifications Comparison of MB-OFDM and IR DS-UWB techniques for WPAN

Specifications	MB-OFDM	IR (DS-UWB)
Number of Sub-bands	3 mandatory, up to 14	2 (3.1–4.85 GHz and 6.2–9.7 GHz)
Sub-band Bandwidth	528 MHz	1.75 GHz (lower band) 3.5 GHz (higher band)
Number of Sub-carriers	122	No sub-carriers (baseband signals)
Spreading Factor	1, 2	1–24
Data rates	55, 80, 110, 160, 200, 320, 480 Mbps	28, 55, 110, 220, 500, 660, 1000, 1320 Mbps (lower band)
Modulation	QPSK	BPM (mandatory), MBOK
Multiple Access	Based on time-frequency codes	Based on PN codes

7.4.1 Interference

In interference and coexistence with other radio services, the MB-OFDM gains some advantages because of the possibility to turn off some carriers that may interfere, or to be interfered, with other systems. Due to this spectral flexibility, MB-OFDM may also increase the robustness against frequency selective fading. However, MB-OFDM systems are very sensitive to ICI and if sub-carriers' orthogonality is lost, it can result in performance degradation. In the MB-OFDM the multipath energy not captured during the CP window will also deteriorate the performance.

The spectral densities of IR transmitted signals can be made lower in comparison to MB-OFDM signals, by spreading the information over larger bandwidths. Consequently, for the IR signals, the probability of interception decreases, which is useful for military applications, as well as interference to narrowband victim receivers [17]. Also spreading signals over larger bandwidths increases the immunity to narrowband interference and ensures good multiple access capabilities [18], [19]. By using wider bandwidths, the IR-UWB fine time resolution characteristic may be used to mitigate effects of fading [20], and possibly solve signal replicas due to multipath. IR systems have, however, the possibility to suffer from several sources of interference, due to the large bandwidth in consideration.

The IR DS-UWB will minimize interference to other systems, by generating signals with lower power levels than the white noise, not causing harmful interference to existing users in the spectrum, a key concern of worldwide regulators. However, advocators of the MB-OFDM do not support DS-UWB because of possible interference concerns [21]. This is due to the difficulties inherent to IR systems to deal with interferers because of the very wide bandwidth.

The IR TH-UWB signals are very narrow in time and therefore, very difficult to intercept as they are also "noise-like". In this technique to correct demodulate a

TH-UWB signal, the precise synchronization and knowledge of PN code is needed. Increasing the cardinality of the TH code increases the complexity of interception of these signals. Moreover, in the M-ary PPM modulation, it is also necessary to know the exact time shifts which represent the modulated pulse. Like other IR techniques, the TH-UWB will also suffer from the interference from other systems acting in the same band.

7.4.2 Robustness to Multipath

MB-OFDM presents high robustness against multipath because of usage of CP or ZPP in dealing with Inter-Symbol Interference (ISI) and ICI. Nonetheless, in dense multipath environments, when the multipath delay is larger than guard time, orthogonallity may be lost resulting in ICI and ISI.

By using the widest possible bandwidth to produce the shortest possible pulses, the IR DS-UWB technique may provide robust communications in harsh multipath environments. DS-UWB can obtain diversity by exploiting the correlation properties of the spreading sequences to resolve and combine the signal replicas that are received over multiple independently faded paths. During the de-spreading operation, the unwanted narrowband interference is spread throughout the spread spectrum bandwidth, which will reduce its effect on the desired signal. However, a disadvantage of the IR DS-UWB technique is that, in the multipath environments, ISI can severely degrade the performance, as the spreading factor is relatively small for high data rates when compared to that of the traditional DSSS systems [22].

The key motivation for the use of IR TH-UWB is the ability to highly resolve multipath with the possibility of implementing and generating UWB signals with relative low complexity [23]. The TH-UWB systems in conjunction with the T-R method may be very robust in multipath environments, not only due to the short time duration of pulses, but also because of advantages of T-R technique, as explained in Chapter 6, and the inherent advantages of the IR technology. By using wider bandwidths, the IR-UWB fine time resolution characteristic may be used to solve signal replicas due to multipath. However, it should be noted that, in the harsh multipath environments the excess delayed pulses can interfere with other transmitted modulated pulses giving rise to difficulties in the TH-UWB receivers to recover the signal.

7.4.3 Performance

Reference [22] reports similar data rates for the MB-OFDM and IR DS-UWB systems based on the WPAN proposals. The two techniques provide similar performance for the AWGN channel with ideal channel estimation. The MB-OFDM scheme is shown, however, to slightly outperform the DS-UWB. The difference is thought to be a result of the different coding structures employed in the two systems. For the four IEEE 802.15 indoor fading channels, the MB-OFDM system generally outperforms the DS-UWB when hard Decision Feedback Equalizer (DFE) is used for the DS-UWB system. However, this trend is reversed in some cases when the Matched Filter Bound (MFB) is considered for the DS-UWB system. Hence, the performance comparison of the two systems depends critically on the assumptions made on the DS-UWB equalizer.

Performance of a system based on the MB-OFDM WPANs proposal, over four IEEE 802.15 indoor channels is reported in [24]. Channel estimation is done by implementing a least-square error (LSE) channel estimator. LSE estimator is chosen against the perfect Channel State Information. A simple least-squares error channel estimator has shown a performance within 0.5–0.7 dB of the perfect channel state information case, for the proposed MB-OFDM system.

According to [25], MB-OFDM requires a higher SNR to achieve the same BER comparing to IR DS-UWB.

The BER performance of TH-UWB and DS-UWB systems in high data rate environment (100 Mbps) is reported in [26]. It is demonstrated that DS-UWB outperforms TH-UWB for a number of users higher than 10, due to the larger spreading factor.

7.4.4 Complexity

MB-OFDM may have some advantages as the OFDM technique has already been implemented in other communication systems, although with much narrower bandwidth comparing to the UWB bandwidth. MB-OFDM can capture multipath energy with a single RF chain and simplified synthesizer architectures may relax band switching time requirements. The longer duration of the OFDM symbol makes it much less sensitive to timing synchronization error than IR DS-UWB. The robustness against multipath eliminates the need of using a complex equalizer [22]. Also, by comparing to the IR technique, MB-OFDM systems do not need to process an extremely large bandwidth.

MB-OFDM is, however, more sensitive to the frequency offset and RF phase noise due to its narrowband sub-carriers. Supporters of DS-UWB criticize the MB-OFDM systems, because of their complexity which results from using complex FFT, when compared to the IR [21]. The complexity of the OFDM system in MB-OFDM varies logarithmically with the FFT size [6] and thus MB-OFDM systems require relatively large computational power due to the use of FFT [22]. Also high PAPR and difficult synchronization are problems in OFDM that may lead to the increase of complexity of the transceivers.

In the IR, the complexity in antennas design and challenges in building RF circuits, such as achieving high-speed analog-to-digital and digital-to-analog converters, and processing a very large bandwidth, are difficult issues. IR systems are normally simple when looking at the transceiver block diagrams. The IR transceivers may be baseband, avoiding Intermediate-Frequency (IF) stages. However, digital complexity to capture multipath energy in dense multipath environments may be difficult to accomplish [27]. Comparing to MB-OFDM, IR techniques need more complex timing synchronization to achieve an acceptable performance. The impulse radio has more resolvable multipath components because of its wider bandwidth and therefore may need to use a RAKE-receiver, [28]. An equalizer may also be needed in the DS-UWB receiver structure, due to the time dispersive nature of the channel, that results in ISI and consequently, in performance degradation [6], [22]. IR receivers can use the T-R technique, with possible improvement of performance and may avoid stringent synchronization requirements and lower the complexity of the IR receiver. A simple T-R receiver may therefore, gather the energy from many resolvable multipath avoiding the use of RAKE-receivers and making the IR receiver much simpler. Other techniques such as interference cancellation can also be used to deal with a very wide bandwidth [29], with the complexity trade-off.

It should be mentioned that the implementation of third generation mobile systems based on Wideband Code Division Multiple Access (WCDMA) can be useful for the successful realization of IR DS-UWB technology. Just like the DS-CDMA technique, the near-far effect is also a problem in the DS-UWB. Solving this problem will increase the complexity of the transceivers.

TH-UWB may provide robust and reliable systems with simple transceiver architectures. T-R receivers represent a low complexity alternative to RAKE-receivers. In the TH technique, transceivers need accurate timing and precise synchronization, which in multipath environments can lead to errors and added complexity? However, with the T-R method, this problem can be minimized and receivers will be much simpler. Meanwhile, the high processing speed needed to generate and correctly demodulate the transmitted pulses is another factor which should be considered in the system's complexity comparison.

7.4.5 Achievable Range-Data Rate Performance

Achievable range-data rate performance, for the MB-OFDM, and IR (DS-UWB and TH-UWB), may be found in [4], [6], [17], [22], [25], [16], [30]-[33].

MB-OFDM is an efficient system, however it does not have the possibility to transmit at so higher rates as the DS-UWB can be seen theoretically in both proposals [4] and [16]. So, the IR DS-UWB proposal has a better scalability than MB-OFDM. Achievable data rates are however different from theoretical ones.

The achievable ranges for MB-OFDM for four IEEE 802.15 channels are presented in Table 7.5, [6]. Similar ranges are also reported in [25].

In Table 7.6 achievable ranges (in meters) for the MB-OFDM and DS-UWB techniques and for different data rates are shown. According to [22], for the

Table 7.5 Achievable ranges (in meters) for MB-OFDM [6]

Bit-rate (Mbps)	MB-OFDM		
	110	200	480
AWGN	20.5	14.1	8.9
CM1	11.4	6.9	2.9
CM2	10.7	6.3	2.6
CM3	11.5	6.8	not applicable
CM4	10.9	4.7	not applicable

Table 7.6 Achievable ranges (in meters) for MB-OFDM and DS-UWB for different data rates [22]

Technology	MB-OFDM			DS-UWB		
Bit-rate (Mbps)	110	200	480	110	220	500
AWGN	20.9	14.2	8.0	19.5	13.8	7.3
CM1	12.2	7.8	2.6	10.1	6.1	N/A*
CM2	12.1	7.2	2.6	9.3	5.7	N/A
CM3	12.6	7.3	N/A	9.4	4.7	N/A
CM4	13.8	8.0	N/A	9.2	N/A	N/A

* N/A: not applicable

AWGN channels, and for the same data rates, almost similar ranges may be reached for the MB-OFDM systems when compared to the DS-UWB. For the four IEEE 802.15 indoor fading channels, the MB-OFDM system generally outperforms the DS-UWB when DFE is used.

This trend is reversed and higher ranges are reached for the DS-UWB in some cases when the Matched Filter Bound (MFB) is considered for the DS-UWB system. According to [17], TH-UWB may allow data rates of around 110 Mbps be transmitted over distances of 15 m in the AWGN channels, and 4–7 m in the multipath channels.

Table 7.7 Comparison summary of UWB techniques

		Interference	Robustness to multipath	Performance	Complexity	Achievable Range-Data Rate
Multiband	MB-OFDM	+++	++	+++	+	+++
IR	DS-UWB	++	++	+++	++	+++
	TH-UWB	++	++	++	++	++

Legend levels: Excellent: +++
Very Good: ++
Good: +
Not Good: –

In [32] the range-data rate compromise of the IR technology is evaluated for the indoor environments, where it is shown that IR can be a good candidate for the high-rate transmission over short ranges.

A summary of the comparison of MB-OFDM and IR technologies is illustrated in Table 7.7.

Problems

Problem 7.1 What are the differences between DS-CDMA and DS-UWB? What do you think of the bit error rate performances of these two systems in the AWGN and Rayleigh fading channels?

Problem 7.2 Using the free space path loss model and assuming same data rates, bit error rates, and achieved ranges of the MB-OFDM and DS-UWB techniques, obtain the required transmit power of two technologies and compare them with each other.

Problem 7.3 What is the effect of pulse shape of the IR technique on the performance and complexity of the UWB system?

Problem 7.4 In this problem we study the optimal use of the OFDM sub-channels through efficient allocation of power and data rates. Each sub-channel of an MB-OFDM system, similar to any other communication systems, is associated with the two fundamental parameters, namely, transmitter power and bit rate (bandwidth). We are interested in maximizing the total data rate (b_{tot}) subject to a total power constraint (complying with the FCC limit):

$$\max b_{total} = \sum_{i=0}^{M-1} b_i$$

$$\text{subject to} \sum_{i=0}^{M-1} P_i = \text{constant}$$

Where P_i, b_i are the sub-channels powers and bit rates allocated to each sub-channel, respectively, and M is the number of subcarriers. For a MB-OFDM transmission with 4 carriers and BPSK modulation through a 2-path (cluster) UWB channel with impulse response

$$h(t) = \delta(t) + a\delta(t - \Delta)$$

where $\Delta = 100$ nsec and $a = 0.5$, and for a fixed total transmit power, obtain the optimal bit rate loading of each subcarrier.

References

1. K. Siwiak and D. McKeown, Ultra-Wideband Radio Technology, John Wiley, Chichester, UK, 2004.
2. M. Ghavami, L.B. Michael and R. Kohno, Ultra Wideband Signals and Systems in Communications Engineering, Wiley, 2004.
3. V.S. Somayazula et al., "Design challenges for very high data rate UWB systems," Proc. Asilomar Conf. November 2002.
4. A. Batra et al., "Multi-Band OFDM Physical Layer Proposal for IEEE 802.15 Task Group 3a," Doc. IEEE P802.15-03/268r3, Mar. 2004; Online resource: http://www.ieee802.org/15/pub/2003/Jul03/03268r3P802-15_TG3a-Multi-band-CFP-Document.doc
5. A. Batra et al., Multi-band OFDM Physical Layer Proposal, Doc. IEEE 802.15-03/267r5, July 2003.
6. A. Batra et al., "Design of a Multiband OFDM system for realistic UWB channel environments," IEEE Transactions on Microwave Theory and Techniques, vol. 52, no. 9, Part 1, September 2004, pp. 2123–2138.
7. B. Muquet et al., "Cyclic prefix or zero padding for wireless multicarrier transmission?," IEEE Transactions on Communications, vol. 50, no. 12, December 2002.
8. A. Batra et al., "Multiband OFDM: A new approach for UWB," ISCAS Conference 2004, pp. V–365–368.
9. G. Wade et al., "Peak-to-average power reduction for OFDM schemes by selective scrambling," Electronic Letters, vol. 32, no. 21, 1996, pp. 1963–1964.
10. R. Fisher et al., "Reducing the peak-to-average power ratio of Multicarrier modulation by selected mapping," Electronic Letters, vol. 32, no. 22, 1996.
11. S. Muller and J. Hubber, "OFDM with reduced peak-to-average power ratio by optimum combination of partial transmit sequences," Electronic Letters, vol. 33, no. 5, 1997.
12. G. Stuber and D. Kim, "Clipping noise mitigation for OFDM by decision aided reconstruction," IEEE Communication Letters, vol. 3, no. 1, 1999.
13. H. Nikookar and R. Prasad, "Weighted OFDM for wireless multipath channels," IEICE Transactions on Communications, vol. E83-B, no. 8, August 2000, pp. 1864–1872.
14. Ultrawideband: High Speed, Short Range Technology with Far-reaching Effects, MBOA-SIG White Paper, September 2004.
15. www.multibandofdm.org
16. R. Fisher and R. Kohno, "DS-UWB Physical Layer Submission to 802.15 Task Group 3a," IEEE P802.15-04/0137r3, July 2004. Online resource: http://www.ieee802.org/15/pub/2004/15-04-0137-03-003a-merger2-proposal-ds-uwb-update.doc
17. A. Molisch et al., "A low-cost Time-Hopping Impulse Radio system for high data rate transmission," EURASIP Journal on Applied Signal Processing 2005, pp. 397–412, 2005.
18. L. Zhao and A. Haimovich, "Performance of Ultra-Wideband communications in the presence of interference," IEEE Journal in Selected Areas in Communications, vol. 20, no. 9, December 2002, pp. 1684.
19. F. Ramirez-Mireles, "Performance of Ultrawideband SSMA using Time Hopping and M-ary PPM," IEEE Journal in Selected Areas in Communications, vol. 19, no. 6, June 2001, pp. 1186–1196.
20. M.Z. Win and R.A. Scholtz, "On the energy capture of Ultra -Wide bandwidth signals in dense multipath environments," IEEE Communication Letters, vol. 2, September 1998, pp. 245–247.
21. F. Nekoogar, Ultra-Wideband Communications: Fundamentals and Applications, Prentice-Hall, NJ, 2006.
22. O. Shin, S. Ghassemzadeh, L. Greenstein and V. Tarokh, "Performance evaluation of MB-OFDM and DS-UWB systems for Wireless Personal Area Networks," 2005 IEEE International Conference on Ultra-Wideband, 05–08 September 2005, pp. 214–219.

23. M.Z. Win and R.A. Scholtz, "Ultra-Wide bandwidth Time-Hopping Spread-Spectrum Impulse Radio for wireless multiple–access communications," IEEE Transactions on Communications, vol. 48, April 2000, pp. 679–691.
24. C. Snow, L. Lampe and R. Schober, "Performance analysis of Multiband OFDM for UWB communication," 2005 IEEE International Conference on Communications, vol. 4, 16–20 May 2005, pp. 2573–2578.
25. M. Welborn, "First principles analysis UWB DS-CDMA and UWB-OFDM in multipath," Doc. IEEE P802.15-03/344r1, September 2003 Online resource: http://grouper.ieee.org/groups/802/15/pub/03/15-03-0344-01-003a-first-principles-analysis-uwb-ds-cdma-and-uwb-ofdm-in-multipath.ppt#1
26. E. Cano and S. McGrath, "TH-UWB and DS-UWB in lognormal fading channel and 802.11a interference," 15th IEEE International Symposium on Personal, Indoor and Mobile Radio Communications, vol. 4, 5–8 September 2004, pp. 2978–2982.
27. H. Nikookar, Ultra Wide Band for Wireless Communication, Short Course in European Conference on Wireless Technology, Paris, 3–7 October 2005.
28. A. Ghorashi et al., "An overview of MB-UWB OFDM," IEE Seminar on Ultra Wideband Communications Technologies and System Design, 8 July 2004, pp. 107–110.
29. S. Han, C. Woo, D. Hong, "UWB interference cancellation receiver in dense multipath fading channel," 2004 IEEE 59th Vehicular Technology Conference, vol. 2, 17–19 May 2004, pp. 1233–1236.
30. M. Benedetto and G. Giancola, Understanding Ultra Wide Band Radio Fundamentals, Prentice Hall, 2004.
31. J. Foerster et al., "Ultra-wideband technology for short- or medium-range wireless communications," Intel Technology Journal, pp. 1–11, 2nd Quarter, 2001. Online resource: http://www.intel.com/technology/itj/q22001/pdf/art_4.pdf
32. H. Sheng et al., "On the spectral and power requirements for Ultra-Wideband transmission," IEEE International Conference on Communications (ICC), vol. 1, May 2003, pp. 738–742.
33. M. Welborn, "Scaling-TG3a-PHY-proposals-for-high-aggregate-data-rates," IEEE P802.15-04/164r1, March 2004 Online resource: http://www.uwbforum.org/documents/15-04-0164-01-003a-scaling-tg3a-phy-proposals-high-aggregate-data-rates.ppt

Chapter 8
UWB Wireless Locationing

Contents

In wireless communications there are several applications for locationing techniques. For example in emergency services like the US E-911, short for Enhanced 911, a locationing technology used in the United States that enables mobile devices to process 911 emergency calls and enable emergency services to locate the geographic position of the caller, or the Europe E-112, the European emergency services focused on the single emergency call number 112. Another example is in location-sensitive billing which enables wireless carriers to offer different rates depending on where the terminal is used and provide the subscribers a new and efficient rate system. By collecting the calling position, the system provider can gradually change the position of base stations to improve the performance. With the location technique it is also possible to improve design and management of wireless systems. Better resource management can be achieved by controlling the transmission power or channel frequencies according to the user position. Another application is in the development of intelligent transportation systems where operators such as fleet owners can use localization techniques to track their vehicle and improve their route plan. Further, ambulance or taxi operators using locationing standards may lower their response time by knowing the position of their vehicles location. Of course navigation for personal use is another possibility.

UWB technology is especially suited for radio locationing because of its very large bandwidth which provides a fine accuracy in ranging. UWB provides low-power and low- cost communication and positioning in one technology. These features allow a new range of applications, including logistics (package tracking), security applications, (localizing authorized persons in high security

H. Nikookar, R. Prasad, *Introduction to Ultra Wideband for Wireless Communications*, Signals and Communication Technology,
DOI 10.1007/978-1-4020-6633-7_8,

areas), medical applications (monitoring patients), family communications/supervision of children, search and rescue (communications with fire fighters, or avalanche/earthquake victims), control of home applications, and military applications [1]. In almost all of these applications locationing is the most important characteristic.

Locationing can be divided into two parts: ranging and positioning [2]. Ranging means computation of the distance between a target node (or point) and the reference node, while positioning means finding the actual location of the target. Locationing systems are classified into three categories based on where the measurements and calculations are performed: network-based, handset-based and hybrid locationing. In the network-based system the location measurements and calculations are done in a central station while in the handset-based configuration these measurements are carried out at the mobile terminal. The hybrid system combines both the network-based and handset-based system. In this scheme the handset measures the distance and relays the information to the central station, where the localization calculations are performed. Several radio positioning techniques have been proposed by exploiting local radio measurements. These techniques are based on one or more measurement types such as angle (AOA – Angle of Arrival), time (TOA – Time Of Arrival) or time difference (TDOA – Time Difference of Arrival) of arrivals and power profiles (RSS – Received Signal Strength). In this chapter we introduce the various radio localization techniques giving significant emphasis to UWB wireless locationing.

8.1 Position Locationing Methods

There are basically three methods to estimate the position: Received Signal Strength (RSS), Angle of Arrival (AOA) and Time of Arrival (TOA) [2]–[3]. These methods are discussed in this section.

8.1.1 Received Signal Strength (RSS)

The signal strength of the mobile (or handset) to be positioned is measured at several stationary receivers. Ideally each measurement provides locationing information on a set of points along the circumference of a circle with radius *D*i, where *D*i is the distance between the *i*th receiver and the object (mobile) to be localized (Fig. 8.1). The position of the mobile is identified as the point where all the circles intersect. Therefore, for 2-dimensional positioning, at least 3 circles are required to solve the ambiguity. Accuracy of the RSS method is usually poor but can be improved by increasing the number of measurements and averaging the results [3]. As this method is based on the received signal power, no synchronization is required between the receivers. However, exact knowledge on the path loss is a vital element for efficient operation of the system.

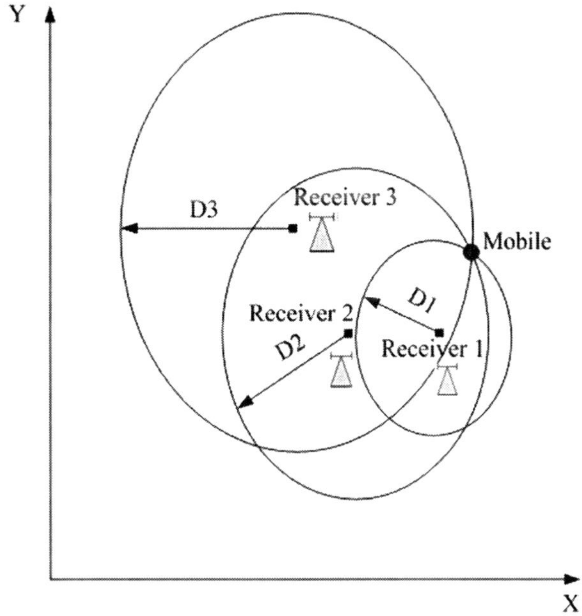

Fig. 8.1 The RSS positioning method

The RSS location techniques are normally feasible for low-cost deployment in GSM networks. The technique is particularly powerful for locating and discriminating indoor users. However, the RSS technique is not suitable for use in multipath fading channels where the variations of the signal strength can be up to 30 dB over a distance of about half a wavelength. This will make the RSS method even more difficult to be used in the DS-UWB systems. Further, in systems such as W-CDMA, where the power is adapted to counter near-far problem, the RSS scheme is untenable.

8.1.2 Angle of Arrival (AOA)

In this method the AOA of the signal sent by the object (mobile) to be positioned is measured at several stationary access points (AP) by steering the main lobe of a directional antenna. As illustrated in Fig. 8.2 each measurement gives a line from the access point to the object to be positioned. Intersection of two lines is the position of the object. Therefore, in theory only two access points are needed for the 2-D locationing. However, in practice because of the multipath effects more than 2 access points will be required. The AOA technique has the advantage of not requiring path loss information or synchronization of the receivers (access points) and need only two or more reference nodes to estimate the location. This technique does not need the time reference. However, it requires a complex hardware for accurate array calibration. It also suffers

Fig. 8.2 The AOA positioning method

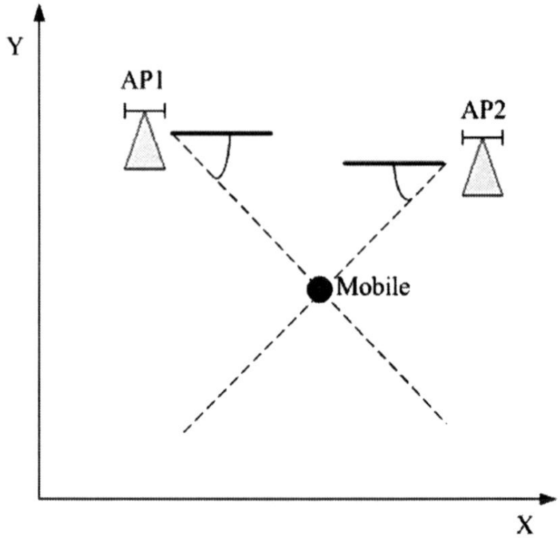

from a large AOA spread in microcells [4]–[5]. Furthermore, in the absence of line of sight (LOS) or when the distance is large the system will yield a poor performance.

8.1.3 Time of Arrival (TOA)

In this location technique the time of arrival of the signal sent by the mobile to be positioned is measured at each receiver (access point). The propagation time of each signal is known and is proportional to the distance. As shown in Fig. 8.3 the measured time provides information in a set of points around the circumference of a circle having the radius of distance between the object (mobile) and the access point. The intersection of the circles is the mobile's position. Similar to the RSS method, for the 2-D location based on the TOA technique, at least three access points are required. Let t_1,t_2 and t_3 denote the flight time of the transmitted signal of the object (mobile) to be positioned, to the respective receivers (access points). The access points 1, 2 and 3 are respectively positioned at locations (0,0), (0,y_2) and (x_3,y_3). Let (x,y) denote the coordinates of the handset (object) to be positioned. By estimating the time of arrivals, the following set of equations should be solved to obtain the position (x,y) of the mobile devices.

$$\begin{aligned} d_1 &= ct_1 = \sqrt{x^2 + y^2} \\ d_2 &= ct_2 = \sqrt{x^2 + (y - y_3)^2} \\ d_3 &= ct_3 = \sqrt{(x - x_3)^2 + (y - y_3)^2} \end{aligned} \tag{8.1}$$

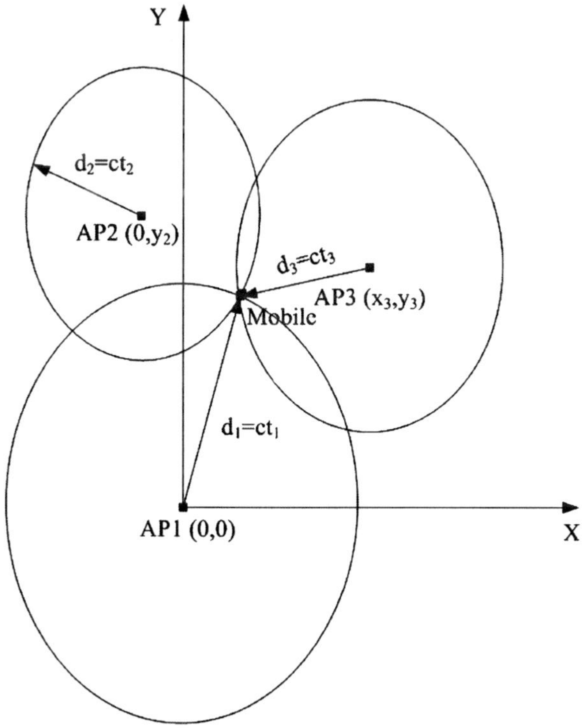

Fig. 8.3 The TOA positioning method

It should be mentioned that to avoid ambiguity in identifying the intersection of circles in Fig. 8.3 all three equations in (8.1) must be considered.

The key factor in time based distance estimator is the arrival time of the first path. TOA utilizes the time delay to get the distance and looks for the intersection of at least three circles to estimate the location. It requires synchronization at both transmitter and receiver side, which is always an important issue in the wireless network, as well as the information of transmission time. If the system is not synchronized or if there is an offset in the time of transmission, the TOA methods cannot work properly. For instance, even a 1 μ sec inaccuracy in the TOA estimation can cause an error of up to 300 m in ranging!

A variant of TOA, the time difference of arrival (TDOA) scheme, can improve the situation. It calculates the target mobile's position according to the time differences between each measurements, rather than the time measurement itself as in TOA. Therefore, as shown in Fig. 8.4, the TDOA searches the hyperbolic intersection and only needs the receiver clock synchronization without the information of time of transmission. The equations for the TDOA positioning become:

$$\sqrt{(x-x_i)^2+(y-y_i)^2}-\sqrt{(x-x_j)^2+(y-y_j)^2}=d_i-d_j, \quad i=1,2,3 \quad \text{and} \quad j=1,2,3 \tag{8.2}$$

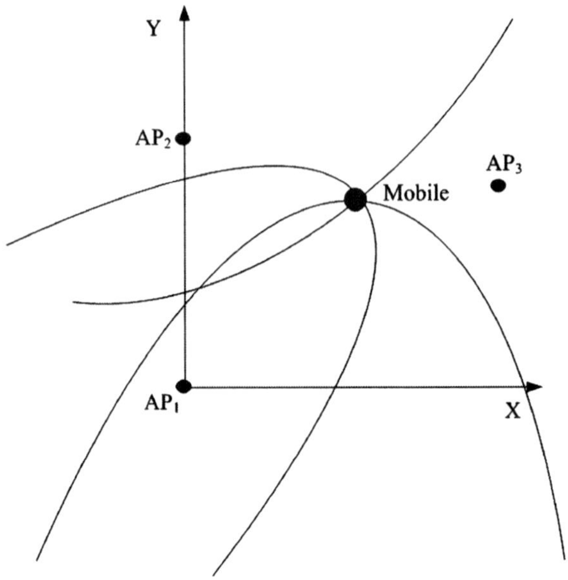

Fig. 8.4 The TDOA positioning method

Efficient methods to solve these nonlinear equations have been reported [6].

Because of the very large bandwidth of the UWB and its fine time resolution, the TDOA and TOA methods are most suited to be used for the positioning with the UWB technique. Theoretically, in the absence of multipath, the variance of TOA estimate is related to bandwidth and the SNR at the receiver [2] as:

$$\sigma^2_{\hat{\tau}} = \frac{N_0}{\int_0^{\infty} 8\pi^2 f^2 |P(f)|^2 \, df} \tag{8.3}$$

where N_0 is the spectral density height of AWGN noise, and $P(f)$ is the spectrum of the UWB pulse. As UWB signals operate under extremely large bandwidths the variance of TOA estimation, as given in Equation (8.3), can become small thereby increasing the accuracy of measurements.

As an example for a UWB pulse having the 2nd derivate of Gaussian shape with spread of 130 psec and exploiting the FCC mask of 3.1–10.6 GHz and assuming $N_0 = 2\times10^{-20}$ W/Hz, the rms of range estimate is obtained as $\sigma_{\hat{\tau}} = 2.5$ mm! This range error is not realistic.

In real time communication, such errors are due to multipath propagation which greatly influences the accuracy of TOA and TDOA positioning methods. Besides multipath effects, the receiver hardware, efficiency in generating the transmission pulses, and multiuser interference affect the decision on range and delay estimation.

It should be mentioned that the combination of the RSS measurements with the TDOA can also be used for locationing. This combination leads to the enhancement of UWB short range positioning with respect to only TOA or TDOA methods [1]–[7].

8.2 Time of Arrival Estimation

As we discussed in the previous section, the RSS method is not an accurate method in locationing, especially in multipath fading environments. The AOA method requires a complex hardware. The TOA and TDOA methods are most appropriate for the UWB positioning because of the fine time resolution of the UWB technology. The major issue in locationing with the TOA or TDOA techniques is the estimation of the arrival time of the UWB pulses. In this section the Time of Arrival estimation techniques will be studied in detail. First four coherent TOA estimation techniques, namely Inverse Filtering, ESPRIT, CLEAN, and Super-Resolution, will be presented followed by a discussion on a non-coherent method.

8.2.1 Inverse Filtering Technique

The inverse filtering method works as following. Assume the transmitted signal is $x(t)$, received signal is $y(t)$ and the channel response is $h(t)$. The received signal can be represented as a convolution of the transmitted signal and the channel impulse response, i.e., $y(t) = x(t) * h(t)$. This relationship in frequency domain is expressed as: $Y(f) = X(f).H(f)$, where $X(f) = F\{x(t)\}$ and $Y(f) = F\{y(t)\}$. Therefore, by dividing $Y(f)$ by $X(f)$, the channel response (and among other channel parameters, the time of arrival of (first) path(s)) can be estimated. This operation in frequency is the dual of that in time domain known as the de-convolution operation.

Figure 8.5 illustrates this procedure. While dividing $Y(f)$ by $X(f)$, great care should be taken as very low signal levels can cause noisy results in the inverse filtering technique. To this end, to avoid noisy results, in Fig. 8.5 only those components that are above –30 dB have been considered in the inverse filtering. In the ranging applications we are interested in the arrival time of the first path, which can be found by searching the position of the first acceptable local maximum according to a threshold in this algorithm. In this method because all the local maxima in the channel are searched in order to find the first arrival, it becomes crucial to choose a proper threshold to distinguish the real delay component and the noise. The threshold are configured based on the maximum value of the channel response, for e.g., Threshold = Peak value – 10 dB.

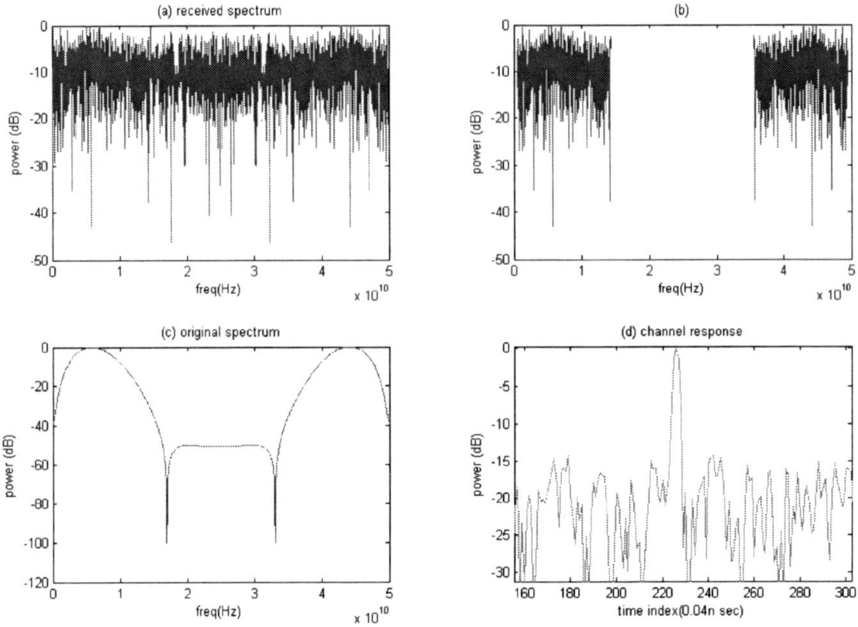

Fig. 8.5 Description of the inverse filtering method. (a) Received signal spectrum, (b) masking the low power components, (c) spectrum of the transmitted signal, (d) channel response

8.2.2 ESPRIT Technique

The ESPRIT (Estimation of Signal Parameters using Rotational Invariance Technique) algorithm is a well-known frequency domain delay estimation method [8]. Let us assume that the original transmitted signal $x(t)$ has L samples $\mathbf{x} = [x(0), x(1), \ldots, x(L\text{-}1)]'$, where $'$ indicates the transpose operation. The received delayed signal $g(t)$ is $\mathbf{g} = [x(0\text{-}\tau), x(1\text{-}\tau), \ldots, x(L\text{-}1\text{-}\tau)]'$, where τ is the delay. The Discrete Fourier Transform (DFT) of $\mathbf{x}$ is written as:

$$\mathbf{X} = \mathbf{F} \cdot \mathbf{x} \tag{8.4}$$

where,

$$\mathbf{F} = \begin{bmatrix} 1 & 1 & \cdots & 1 \\ 1 & W & \cdots & W^{L-1} \\ \cdots & \cdots & \cdots & \cdots \\ 1 & W^{L-1} & \cdots & W^{(L-1)^2} \end{bmatrix}, \quad W = e^{-j\frac{2\pi}{L}} \tag{8.5}$$

and

$$\mathbf{G} = \mathbf{F} \cdot \mathbf{g} = \mathbf{X} \otimes \begin{bmatrix} 1 \\ W^{\tau} \\ (W^{\tau})^2 \\ \cdots \\ (W^{\tau})^{L-1} \end{bmatrix} = diag(\mathbf{X}) \cdot \begin{bmatrix} 1 \\ W^{\tau} \\ (W^{\tau})^2 \\ \cdots \\ (W^{\tau})^{L-1} \end{bmatrix} \tag{8.6}$$

where $\otimes$ denotes the Kronecker multiplication. We compute

$$\mathbf{z} = \{diag(\mathbf{X})\}^{-1} \cdot \mathbf{G} \tag{8.7}$$

Where

$$\mathbf{z} = f(\phi) \tag{8.8}$$

And

$$f(\phi) = \begin{bmatrix} 1 \\ \phi \\ \phi^2 \\ \cdots \\ \phi^{L-1} \end{bmatrix}, \quad \phi = e^{\frac{j2\pi}{L}\tau} \tag{8.9}$$

$$z = \begin{bmatrix} A \\ B \end{bmatrix} = \begin{bmatrix} a \\ a\phi \end{bmatrix}$$

The matrix can be found by extracting the first L-1 components from z to produce A and the last components to form B. After applying pseudo inverse and eigenvalue decomposition the value of ϕ can be found in the eigenvalue results. It is obvious that the delay τ can be derived from ϕ easily.

In the ranging since we are only interested in the arrival time of first delay path, all the delay τ's, which are found in the above-mentioned algorithm, are sorted and the minimum value is taken as the time of arrival. An important point regarding the ESPRIT algorithm is that the number of expected paths needs to be known in advance.

8.2.3 CLEAN Technique

CLEAN is an algorithm that uses the cross-correlation results to find several delay paths [9]–[10]. In this algorithm first a dirty map and a clean map are made. The dirty map is initialized with the received signal and the clean map with zero. Then the correlation between the transmitted signal and the dirty map is obtained. The positions of the peaks of this correlation function are identified. If the value of the correlation function at the peak position is below a predefined threshold the process is stopped otherwise the dirty and clean maps are updated by the value of the transmitted pulse at the estimated time point of the correlation peak. When the process stops the impulse response of the channel is actually the clean map of the algorithm. The TOA is indeed the excess delay of the first path of the impulse response of the channel. If $x(t)$ denotes the transmitted signal and $y(t)$ denotes the received signal, the steps of this algorithm are as follows:

1. Initialize dirty map with $d(t) = Re\{y(t)\}$ and clean map $c(t) = 0$,
2. $\Gamma(t) = \mathrm{Corr}\{x(t), d(t)\}$
3. Find peaks of Γ_i and their positions t_i
4. If $\Gamma_i <$ *Threshold* then stop, otherwise go to Step 5
5. Clean dirty map by $d(t) = d(t) - \Gamma_i\, x(t - t_i)$
6. Update clean map by $c(t) = c(t) + \Gamma_i\, x(t - t_i)$
7. Go to Step 2 until stop
8. The impulse response is $h(t) = c(t)$

It should be noted that through these steps actually a de-convolution in time domain is carried out and the process is akin to inverse filtering. It is also important to notice that in the CLEAN algorithm the whole samples in each pulse repetition intervals are taken as the input signal to the CLEAN algorithm. After applying the CLEAN procedure a series of channel response values will be obtained with different delay times. Then the recorded delay times are searched to find which one is the earliest arrival which will be taken as the first time of arrival. If the period of signal at the input is long enough and the result of CLEAN are sorted to find the time of first arrival, the performance can be definitely improved. Thus, the selection of this period is one of the key factors to control the performance of this procedure. Since the minimum path delay will be chosen regarding to its recorded amplitude, the threshold in the CLEAN procedure is again a remarkable issue. Furthermore, for the CLEAN method it is important that the transmitted signal be known at the receiver side, as it is used in the cross correlation function of the algorithm.

8.2.4 Super-Resolution Technique

Estimation of the arrival time of the pulse, which leads to the ranging, can be carried out by the Super-resolution channel estimation technique. In this technique, high resolution time of arrival is estimated using the frequency domain. Consider a UWB multipath channel with impulse response $h(t)$ as:

$$h(t) = \sum_{m=1}^{M} a_m \delta(t - \tau_m) \tag{8.10}$$

where a_m and τ_m are the amplitude and delay of the mth path, respectively, and M is the number of multipath components. If $x(t)$ denotes the transmitted UWB pulse, the received signal $y(t)$ is

$$y(t) = x(t)^{*}h(t) + n(t) = \sum_{m=1}^{M} a_m x(t - \tau_m) + n(t) \tag{8.11}$$

where $n(t)$ denotes receiver noise. The Fourier transform of the received signal is:

$$Y(f) = F\{y(t)\} = \sum_{m=1}^{M} a_m X(f) e^{-j2\pi f \tau_m} + N(f) \tag{8.12}$$

Where $X(f)$ and $N(f)$ are the Fourier transform of $x(t)$ and $n(t)$, respectively. By sampling the $Y(f)$ at frequencies f_i we have:

$$Y(f_1) = X(f_i) \sum_{m=1}^{M} a_m e^{-j2\pi f_i \tau_m} + N(f_i) \tag{8.13}$$

As can be seen from (8.13) the unknown time delays appear in the complex frequencies, and propagation coefficients as weights. Therefore, the time estimation problem reverts to a classical spectral estimation problem. The Singular Value Decomposition (SVD) or state space approach can be used to estimate the channel parameters. In the following the state space approach is contemplated [11], [12]. Let $Y[m]$, $m \in [-M, M]$, denote the DFT coefficients of $y(t)$ by sampling the received signal with sinc sampling function with the bandwidth of $[-Mf_0, Mf_0]$, where f_0 is the sampling frequency. A matrix $\mathbf{J}$ having the dimension of $P \times Q$ is defined as:

$$\mathbf{J} = \begin{bmatrix} Y_s[0] & Y_s[1] & \ldots & Y_s[Q] \\ Y_s[1] & Y_s[2] & \ldots & Y_s[Q+1] \\ & & \ldots & \\ Y_s[P] & Y_s[P+1] & \ldots & Y_s[P+Q+1] \end{bmatrix} \tag{8.14}$$

where $Y_s[m] = Y[m]/X[m]$ and $P, Q \geq M$. The time delays of the channel are estimated by the SVD decomposition of the Matrix $\mathbf{J}$ and the eigenvalues of the matrix $\mathbf{Z}$ as described below:

$$\mathbf{J} = \mathbf{U\Lambda V}^{H} + \mathbf{U}'\Lambda'\mathbf{V}'^{H} \tag{8.15}$$

where $'$ and ^{H} denote the transpose and Hermition operations, respectively. $\mathbf{U}$ and $\mathbf{V}$ are unitary matrixes and Λ is a diagonal matrix.

$$\mathbf{Z} = \{\mathbf{V}\}^{+}\ \{\{\mathbf{V}\}\} \tag{8.16}$$

Where {.} and {{.}} denote the operation of omitting the first row and last row of (.), respectively, and $(.)^{+}$ denotes the pseudo-inverse of (.).

In this algorithm the poles $z_m = e^{-j2\pi f_0 \tau_m}$ are the eigenvalues of the matrix **Z** and accordingly the time of arrival is estimated.

A major computational requirement of this algorithm is the SVD operation which results in the overall computational order of M^3. An alternative approach is to avoid SVD and estimate the channel parameters using annihilating filter as described below. Let us assume that the $X(f_i)$ can be approximated with a polynomial of degree K, i.e.,

$$X(f_i) = \sum_{k=0}^{K-1} b_k i^k \tag{8.17}$$

where b_k's are the polynomial coefficients. Inserting (8.17) in (8.13) and denoting $a_m b_k = c_{m,k}$, (8.17) can be written as:

$$Y(f_i) = \sum_{k=0}^{K-1} \sum_{m=1}^{M} c_{m,k} i^k e^{-j2\pi f_i \tau_m} + N(f_i) \tag{8.18}$$

The annihilating filter for $Y(f_i)$ is [13]:

$$A(z) = (1 - e^{-j2\pi f_0 \tau_m} z^{-1} K) \tag{8.19}$$

Which means that each exponentials in $Y(f_i)$ of (8.18) is zeroed out by one root of $A(z)$. The time of arrival estimation based on this technique can be summarized as follows:

- Calculate the Fourier transform of the received signal, i.e., $Y[i]$
- Form the correlation matrix of $Y[i]$
- Calculate the null space of the correlation matrix to get path delay estimates (among them the first path TOA)

The computational complexity of the annihilating filter method is M^2, however, it should be noted that this method uses the polynomial root finding which has less numerical precision in the presence of noise. The annihilating filter method for the delay estimation needs low sampling rate [13] which corresponds to lower power consumption of the algorithm. Robustness of this technique to fading is also reported in [13].

8.2.5 Non-Coherent Technique

To avoid sampling rates required by the previous methods, we can use an algorithm based on the energy detection of the first component of the received

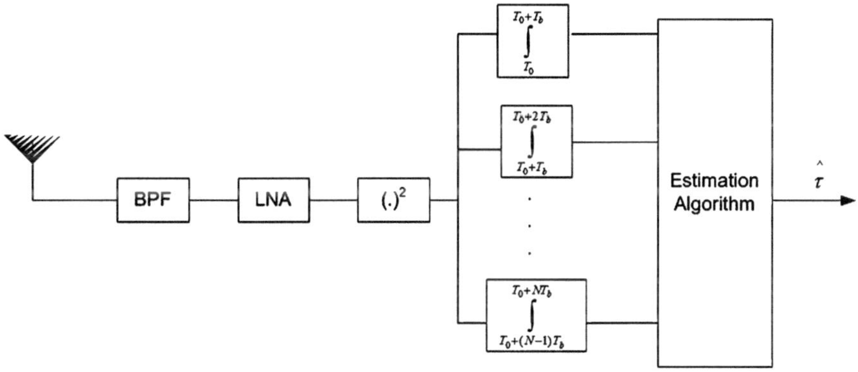

Fig. 8.6 Block diagram of Non-coherent TOA estimation technique

signal together with an analog receiver depicted in Fig. 8.6. The observation window is divided into $N = T_w/T_b$ bins, where T_b is the bin size and T_w is the acquisition time window. To detect the time of arrival of the first path, a bank of N integrators is used. Each integrator calculates the energy of one of the N bins every T_w seconds. The center of the first bin whose energy exceeds a given threshold is selected as an estimation of the time of arrival the first path. With this method there is no need of sampling at the Nyquist rate since the integrators work in analog domain and processing requirements are largely relaxed [14].

8.3 NLOS Location Error

In wireless locationing the observations are sometimes corrupted by non-line-of-sight (NLOS) errors due to multipath propagation. This is the most critical error source in ranging and results in poor location accuracy. For example, in the GSM systems the NLOS discrepancies can cause average errors of the range 400–700 m [15]. The NLOS errors are a problem even when using UWB system for locationing in indoor environments. One approach to handle the NLOS ranging error is to use a premise that NLOS always increases the measured distance. Accordingly, while solving the non-linear equations to obtain the position, a constraint can be imposed that the true location must always lie inside the circles of a radius in each iteration [15]–[16]. By doing so, those solutions of each iteration that fail to satisfy the constraint are not taken into consideration. The use of an intermediate variable consisting of the time-varying NLOS error during the estimation can isolate the NLOS observations and can only consider the LOS results for locationing calculation [17]. Computing the standard deviation of range error, also called time history, can distinguish the NLOS base stations because the NLOS introduces a larger standard deviation in the estimation results [18]. In [19] a special residual to determine the

NLOS base stations for TDOA in CDMA is defined. With the knowledge that which base stations are under NLOS, those measurements from the locationing process can be excluded. Other approach [20] gives different weighting matrix to the measured data to get several sets of results. Then a minimum-distance rule is used to pick up the NLOS base station set to calculate the NLOS error in order to update the weighting matrix, which can minimize the NLOS effect. Reference [21] uses a part of the least square estimator as a residual. Because the NLOS measurements have larger residual than the LOS measurement, it is possible to choose the locationing result from "good" groups in which most of measurements are LOS. Furthermore, different weightings based on the residual squares can be applied to the results from "good" groups in order to get better location results.

8.4 Locationing with OFDM

In this Section we study the position location capability of the OFDM technique. In the OFDM method for positioning the TOA or TDOA are measured by exploiting the OFDM MAC frame. In addition to the application of OFDM in the UWB, the OFDM locationing technique is also important for the wireless LAN as it can be integrated in the existing wireless communication systems without significant changes in both mobile terminals and network infrastructure [24]. In the OFDM locationing technique the TOA or TDOA are measured from the OFDM burst signals. Adopting the notations of [22] suppose:

t_0 = time of transmitting broadcast burst at Access point (AP),
t_1 = time of receiving uplink burst from Mobile Terminal (MT),
τ_{10} = offset of uplink phase within MAC frame (known to AP and MT),
τ_{00} = TOA (to be measured).

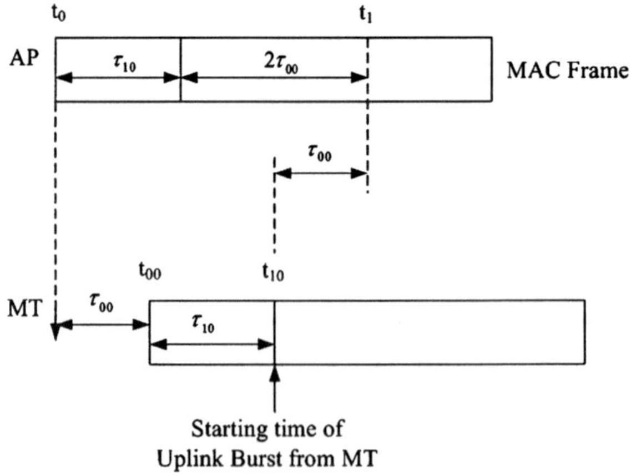

Fig. 8.7 TOA estimation method using OFDM burst

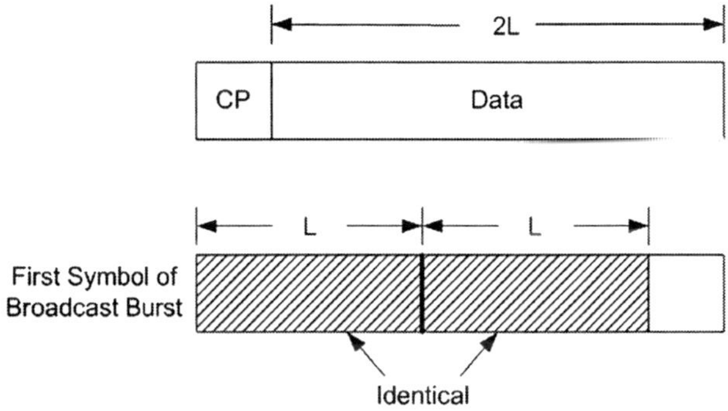

Fig. 8.8 The Hiperlan 2 physical layer burst

According to Fig. 8.7 the TOA from mobile terminal to AP is:

$$\tau_{00} = \frac{1}{2}[(t_1 - t_0) - \tau_{10}] \tag{8.20}$$

In the burst transmission mode receiver continuously scans for the incoming data. The indoor location using OFDM signals in Hiperlan 2 and 802.11 WLANs is reported in [22] and [23], respectively. The Hiperlan 2 physical layer burst has been shown in Fig. 8.8.The MAC frame length of the Hiperlan 2 system is 2 msec.

In Hiperlan 2 the burst preamble consists of training symbols that will be used for the timing synchronization and frequency offset correction. As shown in Fig. 8.8 the first symbol in broadcast burst preamble contains 2 identical parts. The time synchronization is performed by searching for the training symbol with 2 identical halves. A metric M is formed by performing the sliding correlation of 2 consecutive parts of received signal $r(k)$, (which has a length of L), as:

$$\begin{aligned} M(d) &= \frac{|P(d)|^2}{[R(d)]^2} \\ P(d) &= \sum_{m=0}^{L-1} r^*(d+m).r(d+m+L) \\ R(d) &= \sum_{m=0}^{L-1} |r(a+m+L)| \end{aligned} \tag{8.21}$$

The first peak of the sliding correlation indicates the starting point of the first symbol and the last one shows the starting point of the second sample. Simulation results of the technique for the Hiperlan 2 (with 52 carriers, subcarrier frequency spacing = 0.3125 MHz, sampling rate = 20 MHz, 80 samples per symbol, and 16 samples in CP) for the AWGN and an exponentially decaying multipath channel are reported in [22]. For a 5-path channel where the amplitude of each path is calculated from the exponential distribution for large signal to noise ratios a mean range error of 7.5 m is reported [22]. It should be noted that the time synchronization method in (8.21) is a simple one as only one OFDM training symbol is used. By using a more complex timing synchronization method lower mean range error of OFDM technique for location is expected.

8.5 Summary

In wireless communications there are several applications for locationing techniques. UWB technology is especially suited for radio locationing because of its very large bandwidth which provides a fine accuracy in ranging. UWB provides low-power and low- cost communication and positioning in one technology. These features allow a new range of applications, including logistics, security applications, medical applications and military applications.

In this chapter we introduced various radio localization techniques giving significant emphasis to UWB wireless locationing. These techniques are based on one or more measurement types such as angle of arrival), time of arrival or time difference of arrival and received signal strength. Special attention was paid to coherent time of arrival technique (ESPRIT, CLEAN, Inverse Filtering, super resolution) as well as non-coherent method. The problem of multipath propagation and non-line-of-sight location error was explained and the locationing capability of OFDM technique was studied.

Problems

Problem 8.1 Explain why measurement of distance based on phase of the received signal is not a proper method, particularly in the multipath environments.

Problem 8.2 Consider three access points at (0,0), (5 m,12 m) and (13 m,15 m) and the mobile at (10 m,10 m).

Draw the circles corresponding the distances measured from the access points. What is the impact of path loss exponent (n_0) on the circles and their intersections if the path loss exponent is increased from free space ($n_0 = 2$) to 4.

Problem 8.3 In this problem we want to study the effect of UWB pulse shape on the standard deviation of TOA error in an AWGN channel. Consider a Gaussian pulse $p(t)$ as:

$$p(t) = \frac{A}{\sigma\sqrt{2\pi}} e^{-t^2/2\sigma^2} \qquad -\frac{T_p}{2} \leq t \leq \frac{T_p}{2}$$

where T_p is the pulse length, A is the pulse amplitude, selected according to the FCC limit, and σ is the spread of the pulse. Show that the nth derivative of the Gaussian pulse can be written as:

$$P^{(n)}(t) = -\frac{n-1}{\sigma^2} p^{(n-2)}(t) - \frac{1}{\sigma^2} p^{(n-1)}(t)$$

(a) For $n = 2$, $\sigma = 39$ psec, obtain the pulse length T_p to fulfill the FCC mask and sketch the PSD of the pulse.
(b) Repeat part a) for $n = 2$, $\sigma = 116$ psec.
(c) Using the results of part (a) and (b) what is the effect of the pulse bandwidth on the accuracy of TOA estimation (and consequently on the ranging error) in an AWGN channel?
(d) Fulfilling the FCC emission mask, for $n = 2$, 5 and 8, $\sigma = 40$ psec and $T_p = 0.28$ nsec, calculate the range error in an AWGN channel. What is the effect of pulse shape on the ranging error?

References

1. S. Gezici et al., "Localization via Ultra-Wideband radios", IEEE Signal Processing Magazine, July 2005, pp. 70–84.
2. M. Benedetto, Understanding Ultra Wide Band: Radio Fundamentals, Prentice Hall PTR, NJ. 2004.
3. M. Ghavami, L.B. Michael and R. Kohno, Ultra Wideband Signals and Systems in Communications Engineering, Wiley, 2004.
4. J.J. Caffery and G.L. Stuber, "Overview of radio location in CDMA cellular systems", IEEE Communications Magazine, vol. 36., no. 4, April 1998, pp. 38–45.
5. K. Krizman et al., "Wireless position fundamentals, implementation strategies, and source of error", IEEE Vehicular Technology Conference, 1997, pp. 919–923
6. Y.T. Chan and K.C. Ho, "A simple and efficient estimator for hyperbolic location," IEEE Transactions on Signal Processing, vol. 42, no. 8, August 1994, pp. 1905–1915.
7. Z. Sahinoglu and A. Catovic, "A hybrid location estimation scheme for partially synchronized wireless sensor networks", ICC 2004, pp. 3797–3801.
8. R. Roy and T. Kailath, "ESPRIT-Estimation of signal parameters via rotaional invariance techniques", IEEEE Transactions on ASSP, July 1989, pp. 984–995.
9. R.G. Vaughan and N.L. Scott, "Super-resolution of pulsed multi-path channels for delay spread characterization", IEEE Transactions on Communication, March 1999, pp. 343–347.
10. V.S. Somayazula, J.R. Foerster, S. Roy, "Design challenges for very high data rate UWB systems", Proc. Asilomar Conf., November 2002.
11. B.D. Rao and K.S. Arun, "Model based processing of signals: A state space approach", Proc. IEEE, vol. 80, no. 2, February 1992, pp. 283–309.
12. I. Maravic et al., "High resolution acquisition method for wideband communication systems", IEEE ICASSP2003 Conference, pp. IV–133–IV–136.

13. J. Kusuma, I. Maravic and M. Vetterli, "Sampling with finite rate of innovation: Channel and timing estimation for UWB and GPS", IEEE International Conference on Communications, ICC2003, pp. 3540–3544.
14. Z. Irahhauten, G. Bellusci, G. Janssen, H. Nikookar and C. Tiberius, "Investigation of UWB ranging in dense indoor multipath environments", 13th IEEE International Conference on Communication Systems," October 2006, Singapor.
15. J.J. Caffery and G.L. Stuber, "Overview of radio location in CDMA cellular systems", IEEE Communications Magazine, 1998, vol. 36, no. 4, pp. 38–45.
16. X. Wang, Z. Wang and B. O'Dea, "A TOA-Based location algorithm reducing the errors due to non-line-sight propagation", IEEE Transactions on Vehicular Technology, vol. 52, no. 1, January 2003, pp. 112–116.
17. M.P. Wylie and S.S. Wang, "Robust range estimation in the presence of the non-line-of-sight error", 54th IEEE Vehicular Technology Conference, vol. 1, October 2001, pp. 101–105.
18. M.P. Wylie and J. Holtzman, "The non-line-of-sight problem in mobile location estimation", IEEE International Conference on Universal Personal Communications, vol. 2, September 1996, pp. 827–831.
19. L. Cong and W. Zhuang,, "Non-line-sight error mitigation in TDOA mobile location", IEEE Global Telecommunications Conference, November 2001, pp. 680–684.
20. W. Wang, "A new location algorithm for mitigating NLOS propagation errors in microcell envrionment", International Conference on Communication Technology, April 2003, pp. 850–853.
21. P. Chen, "A non-line-of-sight error mitigation algorithm in location estimation", IEEE Wireless Communications and Networking Conference, September 1999, pp. 316–320.
22. X. Li et al., "Indoor geolocation using OFDM signals in Hyperlan 2 wireless LAN", PIMRC 2000, pp. 1449–1453.
23. X. Li, K. Pahlavan, M.L. Aho and M.Ylianttila, "Comparison of indoor geolocation methods in DSSS and OFDM wireless LAN systems", IEEE Vehicular Technology Conference, 2000, pp. 3015–3020.
24. K. Pahlavan and X. Li, "Indoor geolocation science and technology", IEEE Communications Magazine, February 2002, pp. 112–118.

Chapter 9
UWB Applications

Contents

The major characteristics of UWB, i.e., extremely large bandwidth, low power, short-range high data rate communication, robustness against fading, immunity to multipath, multiple access capability, low cost transceivers and precise positioning, motivate several potential applications for this technology. Thus far the UWB technology has been mainly applied to military (especially radar) appliances [1]. In this chapter we study various commercial wireless applications of this technology. The applications have been broadly classified into 6 groups namely – Adhoc Networking, Wireless sensor networks, Radio Frequency Identification or RFID, Consumer Electronics, Locationing and Medical applications (Fig. 9.1). In the following sections we shall briefly explain these applications.

9.1 Wireless Ad hoc Networking

One of the promising applications of UWB technology is in wireless ad hoc networks. Wireless ad hoc networks are networks of nodes (hosts) which are mobile and have no permanent infrastructure. If a fixed or regular infrastructure is available the wireless network is considered as a mesh network. In wireless ad hoc networks multiple hops are used for routing, and the routing changes with time. The major advantages of wireless ad

H. Nikookar, R. Prasad, *Introduction to Ultra Wideband for Wireless Communications*, Signals and Communication Technology,
DOI 10.1007/978-1-4020-6633-7_9,

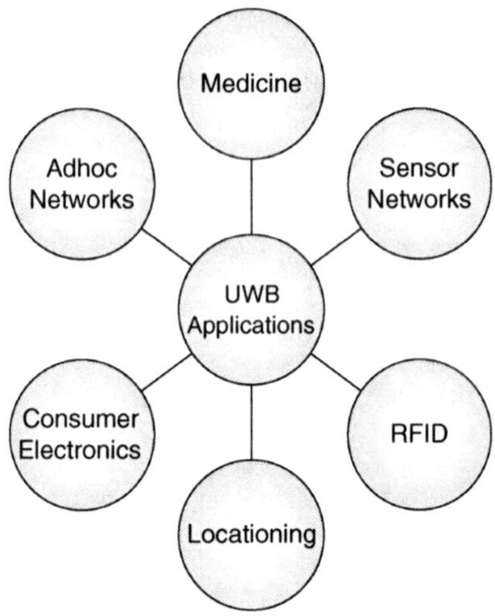

Fig. 9.1 Classification of UWB commercial wireless applications

hoc networks in comparison to traditional communication networks are [2]:

- Easy deployment as no infrastructure is needed,
- Better mobility and flexibility as wireless ad hoc networks can be established or torn down in a very short time,
- Less transmit power for the mobile node because of multi hop, and accordingly lower radio emissions,
- Higher frequency reuse enabling higher capacity coverage,
- Possibility of beyond line of sight (LOS) communication at high frequencies because of multi-hop support,
- More economical as they eliminate fixed infrastructure costs and reduce power consumption of mobile nodes.

For communication between the nodes different radio technologies can be used. For example for the Wireless Personal Area Networks (WPAN) with coverage of up to 20 m the Bluetooth or UWB technology is suggested. For the Wireless Local area Networks (WLAN) with the coverage area of about 100 m, the OFDM or CDMA have been proposed (802.11-a,g and b, respectively). And for Wireless Metropolitan area Networks (WMAN) with the coverage of several kilometers single carrier modulation, OFDM in combination of TDMA, and OFDM are proposed (IEEE 802.16e).

In wireless ad hoc networks each mobile can act as a terminal and as a router. This provides a higher level of autonomy compared to the traditional fixed communication infrastructure. However, the location of mobile, the constraint on power consumption of battery-powered mobile terminals and multipath interference are main concerns in wireless ad hoc networks [3]. UWB technology is an interesting and new application for the wireless ad hoc networks. It can address all of the above-mentioned concerns. The fine time resolution of this technology and the precise locationing capability, low transmit power as well as the robustness against fading makes UWB an ideal emerging radio technology for the wireless ad hoc networks. Among the parameters of UWB wireless ad hoc networks, throughput is a key performance measure of the system. The capability of the UWB technology to locate a node accurately can be used in the routing in order to maximize the throughput [4]–[5]. By optimal allocation of radio resources, such as transmission rate and power, the throughput of UWB wireless ad hoc networks can be maximized [6]. Optimal power control, scheduling and routing in UWB networks are discussed in [7]. New strategies of path selection in the UWB based ad hoc network are reported in the literature [8] where power dependent cost functions are defined and minimized leading to high network performance and low emitted power.

9.2 UWB Wireless Sensor Networks

Wireless sensor networks consist of individual sensor nodes distributed over a given area. The sensors are used to monitor some physical phenomena in the environment (such as temperature, humidity, position, speed, motion, etc.). A unique feature of sensor networks is the cooperative effort of sensor nodes for signal information processing. The UWB can be a remarkable technology for the wireless sensor networks because of its small, low power devices that combine location sensing and wireless communication capabilities. The UWB transceivers and antennas can be very small, low power and low cost. Therefore, there are expectations that wireless sensor networking could likely be a mass market for the UWB [9].

One potential application of UWB wireless sensor networks is the application of UWB in the configuration of smart highways. The network consists of a large number of UWB sensors placed every few (ten) meters along both sides of the road, forming a dense, distributed sensor network. The aim will be to realise smart highways by exploiting UWB technology to wirelessly and cooperatively perform multifunctional tasks: vehicle to vehicle communication, road to vehicle communication and vice versa as well as early warning of potentially dangerous traffic situations, location determination and tracking of vehicles [10]–[13].

Several other intelligent applications can be catered to through smart and safe highway, by the enhancement of the basic positioning and communication functionalities of UWB. For instance at road intersections intelligent traffic lights could decide on the peculiar traffic situations by an ad-hoc network instantaneously formed among all the (smart) vehicles present at that moment in that cross point, instead of functioning purely on timing. This and other issues are being currently investigated with great interest, as is shown by the growing research on Vehicular Ad-hoc Networks, with many research and developments [14]–[16] and standardization efforts [17]–[18].

Effective deployment of sensors nodes and networks requires satisfying a set of design issues. Among them, the most challenging is related to the cost and power requirement of the wireless sensors: both have to be as low as possible. Since wiring costs are eliminated and because no expensive pre-existing infrastructure is required, the cost of a single node is very important to justify the overall cost of the network made by a lot of sensors [19]. UWB technology is well qualified to address this issue.

In sensor networks another important requirement (perhaps more important than the cost) is the energy-awareness for extending the life of networks encompassing a lot of terminals with limited power supply (such as solar cells supported units) [20]. Reducing the energy dissipation is a process that has to involve all layers of the system design process, from the single emitter level to the system level. Low power design at hardware physical layer uses different techniques, resulting in considerable energy saving. UWB appears to be a new promising alternative physical layer technology for wireless sensor networks, having many attractive features. In addition to physical layer, other venues need to be explored as well. In wireless sensor networks the sources of power consumption can be classified into two types: communication and processing. The computational collaboration between nodes comes at the cost of exchanging information between them. The largest opportunity in energy reduction lies in the (physical layer driven) protocol stack design, where a trade off between communication and computation can lead to many orders of magnitude energy efficient solutions [21]–[23]. However, the protocol stack of UWB sensor networks still seems to be not mature and a variety of challenges and issues needs to be better understood [24].

9.3 RFID

Radio Frequency Identification (RFID) is an automatic identification technology, similar to barcode, which uses radio waves to communicate with the target tags. Distinct advantage of RFID is that it does not require line-of-sight operation. The read range of RFID is larger than the barcode. The RFID readers can communicate simultaneously with multiple RFID tags and because of this the RFID reader can capture the contents of entire shipment as it is loaded to into a warehouse. Furthermore, the RFID tags can store more data than bar codes [25]. The key issues in RFID technology are to assure connectivity to the tags, to determine accurate position and to reliably communicate sensor status if needed [26]. RFID has been around for decades [25] and is a mature technology, but for reasons of cost and size, its use has been confined to a narrow set of applications [27]. As UWB can provide good connectivity as well as accurate position identification capability it has excellent prospect for RFID applications. The UWB RFID tags and tag readers can be small and of low-cost. Major challenges of RFID are privacy concerns and security. Current implementations of narrowband RFID rely on digital cryptographic primitives which are motivated by the privacy requirement. Consequently, this increases the overall processing latency, power consumption and silicon area of the RFID tag. Using UWB technology the RFID signal can be spread across the spectrum making RFID secure. Moreover, due its very large bandwidth, UWB has better propagation properties than conventional narrowband communication. Therefore, it is expected that UWB RFID to be useful in environments not conducive for narrowband RFID system operation. Application of UWB RFID in hospital environments has been recently reported [28]. The system allows hospitals to track the status and exact location of patients and staff and essential equipment.

9.4 Consumer Electronics and Personal Computers (PC)

Another application of UWB technology is in the area of consumer electronics and PCs. As was discussed in the previous chapters, UWB is capable of short-range high-speed communication. This characteristic is extremely important for consumer electronics market such as DVD players, Digital Camera, MP3 players, Digital TV's, etc. Potential applications include wireless high-speed transmission of data between DVD players/recorders and (HD)TV and PC's peripherals. This technology eliminates the need for cable connection among consumer electronics devices and subsequently increases the freedom and movement of the user.

Besides wireless transmission, UWB over cable is another area where this technology can be applied. UWB technology can be delivered over cables [29]. This could effectively double the bandwidth available to cable television systems [30]. As coaxial cable is a shielded environment, the FCC restrictions will not apply and there is greater manoeuvrability with regard to the UWB transmission spectrum. Furthermore, the UWB over cable does not interfere with television, high-speed internet, voice or other services already provided by the cable television infrastructure [29]. Early results show data rates of about 2 Gbps over 50—70 m of coaxial cable through repeaters. If successful, UWB could ride over existing infrastructure, with data rates of a few Gbps without interfering with legacy connections [9].

And when UWB sensors are integrated with consumer electronics, the resultant technology can spawn novel household devices that can help realise the vision of smart homes and offices. The key to smart homes is to accurately locate the user position, a requirement that UWB is ideally suited to handle.

9.5 Asset Location

UWB offers communication and identification of location in one technology. The precise location measurement and its usage in the asset management is an important application of the UWB technology and will be a significant market in the coming years. The location capability can also be used to locate personnel, inventory items, and vehicular robots in indoor environments. Development and test of a UWB precise asset location (PAL) is reported in [31]–[32]. The PAL system uses the UWB location capability of UWB technology based on TDOA technique. The estimated accuracy of a few feet in open cargo environments with containers has been achieved [31]. Tests on PAL report that the system works well in open and partially loaded cargo spaces. Unlike narrowband RFID systems, UWB appeared to penetrate large cracks between containers, maintaining localization capability during blockage tests [31].

9.6 Medical Applications

UWB signals are not influenced by clothes or blankets, and can even penetrate human body, walls, ground, ice, mud, and many interesting potential applications for UWB in medicine can be envisioned. Hospitals, Operation theatres, Home health care, intensive care units (ICU), pediatric clinics, rescue operations (to look for heart beat under ruins, or soil, or snow) are few potential areas of application [33]–[34]. Some of the major motivations for using UWB radar and wireless communications in the field of medicine are:

- Non-contact based wireless devices

Operating theatres have to maintain a sterile environment to ward off infection. A common problem is when non-sterile instruments have to be used (for e.g. switching on a light or using a pen to document proceedings) during an operation. Use of non contact equipments and technologies is therefore of utmost importance. UWB can contribute constructively to this requirement. They can be readily customised to build non contact wireless equipments for designing a smart operating theatre where commands can be "wirelessly" issued to perform designated tasks. The ultra wideband nature of UWB also comes in handy as it can avoid causing any interference to other medical appliances.

- UWB Radar

Electromagnetic UWB pulses can "see through" human body and hence can be used in medical imaging. Different tissue and body mass have different reflection indices. UWB signals being short duration pulses can easily exploit the difference in reflection indices to give a clearer picture of the organs, including movements, and that too non-invasively. UWB has been shown to detect the movements of the heart wall by exploiting the difference in reflection magnitudes between the heart muscle and the blood it pushes into the vascular tree [33]. Other organs that cane be probed through UWB are vocal cords, vessels, bowels, heart, lung, chest, bladder and fetus.

- Remote and continuous monitoring

UWB can be a suitable technology for remote monitoring the patients and can replace ultrasound. Patients can be monitored remotely and non-invasively for long periods of time without interruption. A typical example would be remote, non-contact and continuous care of mother and child. The applicability of UWB for RFID is translatable to medical environments to track the status and exact location of patients and staff and essential equipment. Moreover, UWB technology can chip in with sensors for monitoring pulse rate, blood pressure, temperature, life signs and can transmit their data without wire. This will be more comfortable to the patients as well as medical personnel in comparison to conventional wired sensors.

- Low power

UWB devices operate under very low power. The UWB transmission power level is very weak and studies [35]–[36] report that they are safe for humans. As a consequence, the technology can be used for monitoring patients for long periods with low battery power without any harmful side effects.

- Low Cost and low Maintenance

A major attraction for UWB appliances is that they can be built from "off the shelf" electronic devices. (A Philips research center at Eindhoven, The Netherlands, demonstrated UWB transmission using 5 cents coins as antennas!) Cheap, easy to use and scalable UWB medical products can readily be tailored to any requirement and mass produced.

9.7 Summary

The UWB technology can spawn a wide range of interesting wireless applications. In this chapter we listed and gave a brief account of few of the UWB wireless applications. The key properties of UWB that prompt their applicability to diverse requirements are: high data rate communication, robustness against fading, immunity to multipath, multiple access capability, low cost transceivers and accurate positioning.

Problems

Problem 9.1 In this chapter we studied the application of UWB for the wireless ad hoc networks. Which UWB technology (Impulse radio or MB-OFDM) better suits this application? Why?

Problem 9.2 Repeat Problem 9.1 for the UWB wireless sensor system application.

Problem 9.3 Compare the narrow-band RFID with the UWB RFID. What is the major draw back of narrow-band RFID system?

Which of the two systems will perform better in the multi-path environments? Why?
Which system will perform better in a severe interference environment? And which of the two will consume lower power?

Problem 9.4 One problem in the PAL system is that for a precise tag position, a minimum number of receivers (typically 3) should have a direct line of sight path. However, due to the nature of indoor environment, there may be a limited number of such direct transmission paths [32]. For example, walls, machinery, containers and other materials may create signal attenuation or even complete signal blockage. What is the solution to this problem? What do you think about the drawback of your solution?

References

1. D. Taylor, Ultra Wideband Radar Technology, CRC Press, 2001
2. R. Hekmat, "Fundamental properties of wireless mobile ad hoc networks," Ph.D. Dissertation, Delft University of Technology, 2005.
3. M. Ghavami, L.B. Michael and R. Kohno, Ultra Wideband Signals and Systems in Communications Engineering, Wiley, 2004.
4. X. An et al., "Novel location-aided routing scheme for UWB ad hoc networks," Proc. IEEE International Symp. Commun., Information Technology, 2005, pp. 634–637.
5. L.D. Nardis et al., "A position based routing strategy for UWB networks," Proc. IEEE Conf. UWBST 2003, pp. 200–204.
6. C. Zou and Z. Haas, "Optimal resource allocation for UWB wireless ad hoc networks," Proc. IEEE PIMRC 2005, pp. 452–456.
7. B. Radunovic and J.Y. Le Boudec, "Optimal power control, scheduling and routing in UWB networks," IEEE Journal of Select Area of Communications, vol. 22, no.7, pp. 1252–1270, Sept. 2004.
8. L.D. Nardis et al., "UWB Ad hoc networks," Proc. IEEE UWBST2002 Conference, pp. 219–223.
9. F. Nekoogar, Ultra Wideband Communications: Fundamental and Applications, Prentice Hall, NJ, 2006.
10. S. Gezici et al., "Localization via Ultra-Wideband Radios – A look at positioning aspects of future sensor networks," IEEE Signal Processing Magazine, vol. 2, no. 4, July 2005.
11. R. Szewczyk, "UWB: Technology and implications for sensor networks," Presentation, NEST Meeting 08/27/2004, available at: http://www.cs.berkeley.edu/~binetude/NEST/aug27.htm.
12. H. Veenstra et al., "15–27 GHz Pseudo-Noise UWB transmitter for short-range automotive radar in a production SiGe technology," Proceedings of ESSCIRC, Grenoble, France, 2005.
13. J.C. Adams et al., "Ultra-wideband for navigation and navigation," IEEE Proceedings Aerospace Conference, 2001, vol. 2, 10–17 March 2001.
14. "The Network-On-Wheels project" available online: http://www.network-on-wheels.de/
15. "The PReVENT project" available online: http://www.prevent-ip.org/
16. A. Arora et al., "A line in the sand: A wireless sensor network for target detection, classification and tracking," Computer Networks, vol. 46, no. 5, Dec. 2004.
17. IEEE 802.11p Task Group, available online: http://grouper.ieee.org/groups/scc32/dsrc/index.html http://ewh.ieee.org/soc/vts/
18. Workshop on Intelligent Transportation, available online: http://wit.tu-harburg.de/

19. I.F. Akyildiz et al., "A survey on sensor networks," IEEE Communication Magazine, Aug. 2002.
20. L. De Nardis, A.G. Di Benedetto, "Joint Communication, ranging and positioning in low bit rate UWB networks," 2nd Workshop on positioning, navigation and communications and 1st UWB Expert Talk, Hannover, Germany, March 2005.
21. C. E. Jones et al., "A survey of energy efficient network protocols for wireless networks," Wireless Networks, July 2001.
22. J.M. Rabaey et al., "PicoRadio supports ad hocultra-low power wireless networking," IEEE Computer, vol. 33, no. 7, July 2000.
23. E. Shih et al., "Physical layer driven protocol and algorithm design for energy-efficient wireless sensor networks," Proc. of the Seventh Annual ACM/IEEE International Conference on Mobile Computing and Networking, July 2001.
24. K.D. Wong, "Physical layer considerations for wireless sensor networks," IEEE International Conference on Networks, Sensors and Control (ICNSC) 2004, Taipei, Taiwan, March 2004.
25. R. Weinstein, "RFID: A technical overview and its application to enterprise," IEEE IT Pro., May–June 2005, pp. 27–33.
26. F. Dowla, "Long-range Ultra-wideband radio frequency identification, online document: www-eng.llnl.gov/pdfs/dist_sys_sensors-8.pdf
27. S. Pradhan et al., "RFID and sensing in the supply chain: Challenges and opportunities," 2005. online document http://www.hpl.hp.com/techreports/2005/HPL-2005-16.pdf.
28. J. Collins, "Hospitals get Ultra-wideband RFID," RFID Journal, online document: http://www.rfidjournal.com/article/articleview/1088/1/1/
29. K. Siwiak and D. McKeown, Ultra-Wideband Radio Technology, Wiley, 2004.
30. Online document: http://www.pulselink.net/
31. R.J. Fontana and S.J. Gunderson, "Ultra wideband precision asset location system," IEEE Conf. UWBST, 2002, pp. 147–150.
32. R.J. Fontana et al., "Commercialization of an UWB precision asset location system," IEEE Conf. UWBST, 2003, pp. 369–373.
33. Staderini, "UWB Radars in Medicine," available online: http://www.roke.co.uk/download/papers/uwb_research_at_rmr_uwb2003.pdf
34. R. Natalia, "Ultra Wide Band (UWB) and Health Applications," IREAN Research Workshop, Virginia Tech, 2005.
35. J.H. Merritt, J.L. Gel and W.D. Hurt, "Considerations for human exposure standards for fast-rise-time high-peak-power electromagnetic pulses," Aviation, Space, and Environmental Medicine, vol. 66, no. 6 pp. 586–589, June 1995.
36. J.R. Jauchem, R.L. Seaman, H.M. Lehnert, S.P. Mathur, K.L. Ryan, M.R. Frei and W. D. Hurt, "Ultra-wideband electromagnetic pulses: Lack of effects on heart rate and blood pressure during two-minute exposures of rats," Bioelectromagnetics, vol. 19, no. 5, pp. 330–333, 1998.

Chapter 10
UWB Regulation

Contents

As mentioned in other chapters, the UWB systems operate in a very large bandwidth necessitating it to share the spectrum with other users as well as with the existing communication systems and consequently, interferences may occur. Regulation of UWB radio spectrum is therefore necessary to establish a framework where UWB systems can peacefully co-exist with legacy systems. Radio regulations are rules which address the coordination of spectrum access amongst multiple wireless services and applications. Existing regulations which thus far focus only on narrowband radios will therefore have to accommodate the UWB paradigm. In this Chapter we focus on the regulation of the UWB technology and discuss how this new technology in wireless communications shapes the way of spectrum sharing and consequently impacts the decisions of radio regulation bodies. The objective of this chapter is to give the reader a flavor of the activities of various regulatory bodies to facilitate and streamline UWB spectral access.

It should be mentioned at this point that this chapter is by no means a comprehensive and updated document on the UWB regulation efforts, but rather an attempt to indicate the importance of the regulation and standardization of this technology.

H. Nikookar, R. Prasad, *Introduction to Ultra Wideband for Wireless Communications*, Signals and Communication Technology,
DOI 10.1007/978-1-4020-6633-7_10,

10.1 UWB Regulation in US

UWB regulation sets upper bounds on the power that can be radiated at any particular frequency, both within and outside the core band of 3.1–10.6 GHz, and is usually specified as a spectral "mask". The Federal Communications Commission (FCC) in the US has set out such a mask to regulate UWB communication. The release of the mask was preceded by significant efforts by the industry to promote the UWB technology and convince the FCC to allow license free access to spectrum under FCC part 15 regulations [1]. The FCC Part 15 Rules permit the operation of classes of radio frequency devices without the need for a license or frequency coordination. It also attempts to ensure a low probability of unlicensed devices causing harmful interference to other users of the radio spectrum. On 14 February 2002 the US FCC issued a First Report and Order for UWB technology and authorized the commercial deployment of UWB technology, though subject to technological and operational constraints. This followed extensive consultations that led the FCC to conclude: "UWB devices can be permitted to operate on an unlicensed basis without causing harmful interference provided appropriate technical standards and operational restrictions are applied to their use".

The UWB radiation mask defined by FCC has been depicted in Fig. 10.1, and the Effective Isotropic Radiation Power (EIRP) figures are re-produced in Table 10.1. In the figure and table the limits are for indoor and outdoor handheld systems.

The FCC continuously evaluates the UWB technology through tests and measurements and makes necessary amendments. Therefore, the FCC regulations on UWB are expected to evolve with time in the course of developments of future technology.

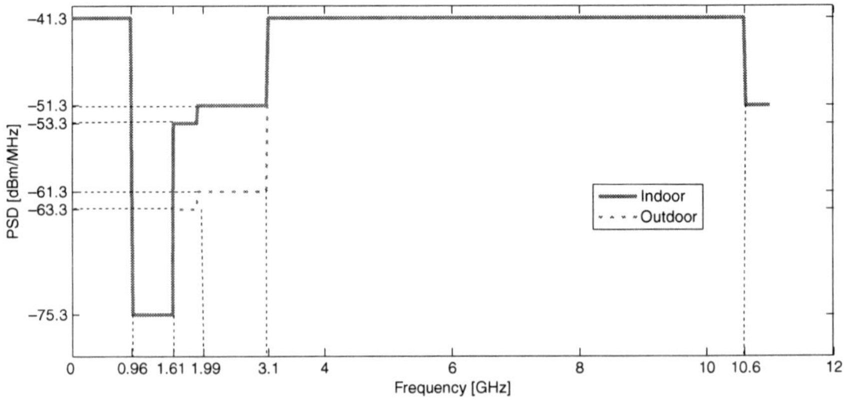

Fig. 10.1 UWB spectral mask as defined by FCC

Table 10.1 FCC limits for indoor and outdoor systems

Frequency Ranges	Indoor EIRP (dBm)	Outdoor EIRP (dBm)
960 MHz–1.61 GHz	–75.3	–75.3
1.61 GHz–1.99 GHz	–53.3	–63.3
1.99 GHz–3.1 GHz	–51.3	–61.3
3.1 GHz–10.6 GHz	–41.3	–41.3
Above 10.6 GHz	–51.3	–51.3

10.2 UWB Regulation in Europe

The organizations involved in the regulation of UWB in Europe are ETSI (European Technical Standard Institute) and CEPT (European Conference of Postal and Telecommunications Administration). These institutions conduct UWB compatibility and spectrum sharing studies and devise regulatory mechanisms. In 2003 the European Union gave a mandate to the ETSI to establish a set of harmonized standards covering UWB applications. Subsequently in 2004 and following the completion of spectrum compatibility studies by CEPT, ETSI established a task group ERM TG31A to develop a set of harmonized standards for short range devices using UWB technology. The UWB standardization working groups include ERM/TG31A covering generic UWB, and ERM/TG31B which covers UWB for automotive applications operating in higher frequency bands, [1], [2]. In June 2005 the European Commission issued its second mandate to CEPT to undertake the technical work required to enable the introduction of ultra-wideband technology in the European Union. The text of the mandates can be found at [3]. In the meeting of March 24th, 2006 held in Oulu, Finland, the Electronic Communications Committee (ECC) of the CEPT adopted a voluntary regulation for UWB which contains the proposed technical elements pursuant to the UWB mandates [4]. The new regulations lay down that UWB equipment should be used predominantly indoors and thus avoid interference. Further, it also imposes a few additional restrictions on device capabilities. For example, it rules that UWB equipments must cease transmission within 10 s unless they receive acknowledgement from an associated transceiver that its transmission is being received. Further, the outdoor use of UWB technology should not include a fixed outdoor location or connected to a fixed outdoor antenna or in vehicles.

The technical requirement for the devices using UWB technology in bands below 10.6 GHz permitted under ECC decision is depicted in Fig. 10.2 and Table 10.2. The limits are for indoor UWB communication. Even while recognizing issues of scalability and conformance to global standards for UWB applications, the regulatory bodies in Europe are more cautious than that of the USA [5]. In order to compare the ECC and FCC limits, both masks are illustrated in Fig. 10.3. From this figure it is seen that the European approach to UWB emission is more restrictive than FCC, and only in the band 6–8.5 GHz does it have the same emission level as the FCC.

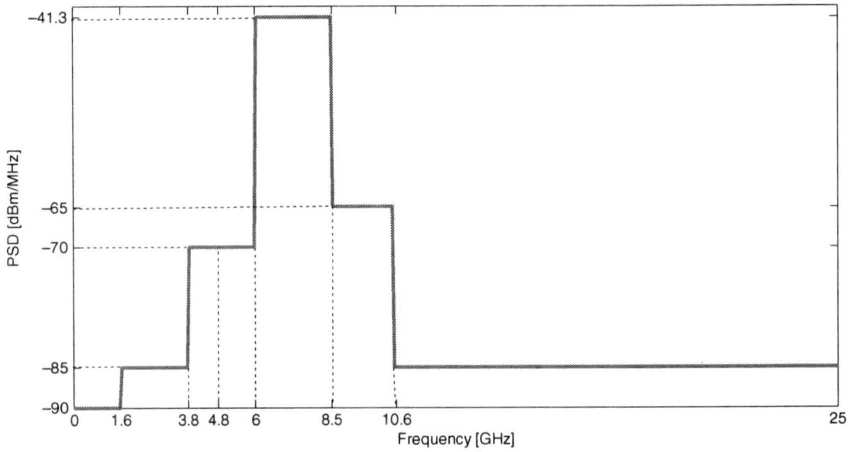

Fig. 10.2 The European spectrum mask

With these recent ECC decisions the UWB technology has gained the green light from the European Commission's Radio Spectrum Committee paving the way for the technology to be legally in the European Union [6].

Of particular interest in Europe is the work carried out by the Pervasive Ultra Wideband Low Spectral Energy Radio Systems (PULSERS) project where UWB transmission at data rates higher than those that the FCC allows has been studied. Importance of this work is its spectral mask imposed to avoid interference with other technologies in band. In the second phase of PULSERS project attention is paid to the chip and module implementation for transceivers with the goal of reducing the cost and power dissipation of higher speed UWB subsystems [7], [8]. The Project PULSERS, entirely focusing on the advancement of UWB radio technology, plays an active role in the European regulatory process leading to adoption of UWB regulations for next generation wireless devices.

Table 10.2 Technical requirements for devices using UWB technology in bands below 10.6 GHz in Europe

Frequency ranges	Mean EIRP density
Below 1.6 GHz	–90 dBm/MHz
1.6–3.8* GHz	–85 dBm/MHz
3.8–4.8* GHz	–70 dBm/MHz
4.8–6 GHz	–70 dBm/MHz
6–8.5 GHz	–41.3 dBm/MHz
8.5–10.6 GHz	–65 dBm/MHz
Above 10.6 GHz	–85 dBm/MHz

*ECC is sill considering whether or not to adopt a separate decision covering the frequency band 3.1–4.8 GHz.

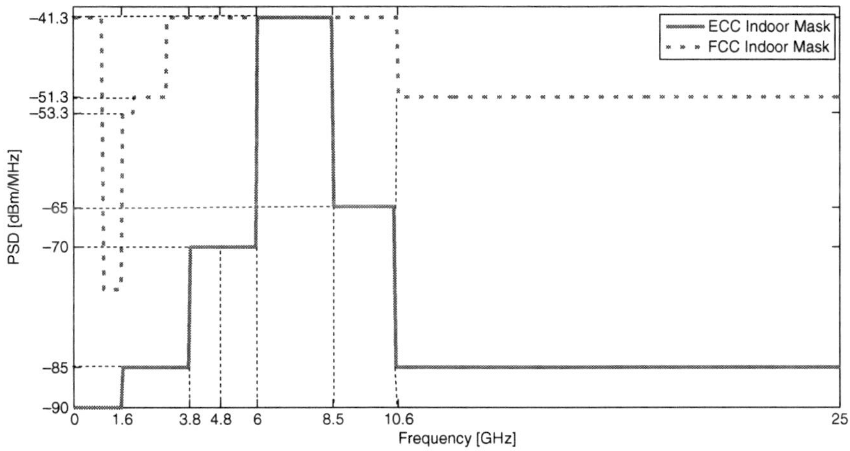

Fig. 10.3 Comparison of the FCC indoor UWB mask with the European one

10.3 UWB Regulation in Japan

The Japanese UWB radiation mask for indoor devices has two bands; from 3.4 to 4.8 GHz and from 7.25 to 10.25 GHz. For the 3.4–4.8 GHz band, it is required to use a technology to reduce interference with other radio services. This interference mitigation is called Detect And Avoidance (DAA) to ascertain the coexistence with incumbent systems and new services such as 4G systems. However, temporary measures are taken until end of 2008 to permit the use of 4.2–4.8 GHz band without an interference reduction technology. It should be noted that no DAA is required for the band 7.25–10.25 GHz. The preliminary Japanese UWB emission mask is depicted in Fig. 10.4. Similar to the FCC

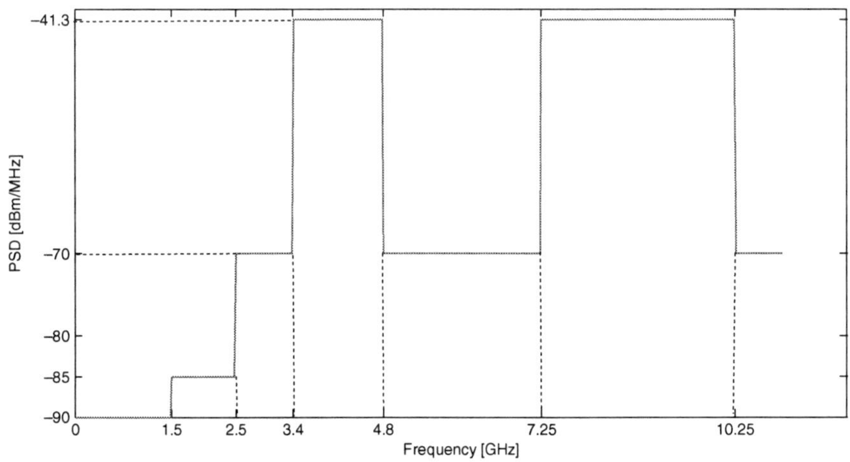

Fig. 10.4 The preliminary UWB Japanese emission mask

mask, the average power spectral density is limited to –41.3 dBm/MHz or lower on both bands. Although the latest announcement of Japan's UWB regulation was limited to the approval of indoor usage, outdoor usage will also be discussed. Japanese authorities also plan to set up a working group to discuss regulations on the 24 GHz band for use in automotive radars [9], [10].

10.4 UWB Regulation in Korea

Korea is working on the regulation of UWB radio through its Reform Frequency Regulation for Unlisenced Stations. From the Korean proposal it is expected that the emission levels in the satellite digital multimedia band and WiBro (Wireless Broadband) operating in the 2.3 GHz, be more restrictive. For the frequency range of 1–10 GHz, the Korean emission level is –66.5 dBm/MHz, which is about 25 dB lower than the FCC limit [9].

10.5 UWB Regulation in Singapore

In Singapore the established UWB Friendly Zone (UFZ) allows test and trial of UWB technology by developers. The emission mask in the frequency range of 2.2–10.6 GHz is –35 dBm/MHz which is 6 dB higher than the FCC limit (i.e., –41.3 dBm/MHz). The Singapore's UWB emission mask is shown in Fig. 10.5.

Because of the importance of the interference from UWB devices on the regional radio communication services, the Asia-Pacific Tele-community has started an activity by forming a subgroup on UWB Spectrum Harmonization

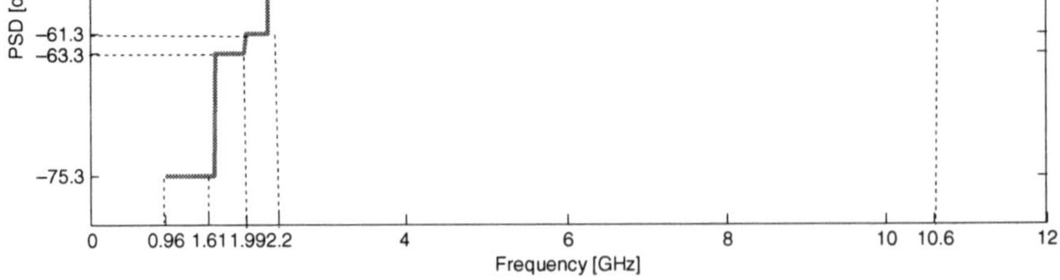

Fig. 10.5 The emission level of Singapore's UWB regulation

Activities for the Asia-Pacific region. The major focus of the work group will be to review the existing policies and spectrum allocations, report on compatibility issues with existing communication services and develop a proposal for the harmonized use of UWB radio technology in the Asia-Pacific region [9], [11].

10.6 UWB Regulation in ITU

In January 2003 the ITU (International Telecommunications Union) formed task group (TG) 1–8 to investigate all UWB issues including compatibility with other radio services. The aim of the task group is to provide a single focal point in dealing with regulatory and technical aspects of UWB. TG1-8 activities are divided into 4 working groups and is responsible for the following outputs [12]

- Recommendation on UWB characteristics
- Recommendation and report on the impact of UWB on other radio systems (ITU-R 227/1)
- Recommendation on a spectrum management framework for UWB (ITU-R 226/1)
- Recommendation on techniques for measuring UWB emissions

Results of technical studies of ITU-TG1-8 are expected to have strong impact on the future regulatory arrangements for UWB in various countries around the world which are monitoring the ITU activities and waiting for the recommended spectrum management framework for the UWB radio technology. Further details, status and recent developments of the UWB regulation in the ITU can be found from the Website of ITU [13].

10.7 IEEE Standardization

In the wireless zone the IEEE standardization activities [14] are IEEE 802 for WLAN (Wireless Local Area Networks 802.11), WPAN (Wireless Personal Area Networks 802.15), WMAN (Wireless Metropolitan Area Networks 802.16), and MBWA (Mobile Broadband Wireless Access networks 802.20). These activities, along with the radio regulatory technical advisory group (TAG) of 802.18 and coexistence TAG with existing standards (802.19), standards for interoperability between heterogeneous network types (802.21) and Wireless Regional Area Networks (802.22) are shown in Fig. 10.6.

Basically the IEEE 802.15 activity focuses on development of standards for Wireless Personal area Networks (WPAN) or short range networks. The 802.15.1 has derived a WPAN standard based on Bluetooth. The 802.15.2 developed recommended practices to facilitate coexistence of WPAN (802.15) with WLAN (802.11). The UWB standardization activity of the IEEE is more concentrated in the 802.15.3 with the task group 3a focusing on UWB physical

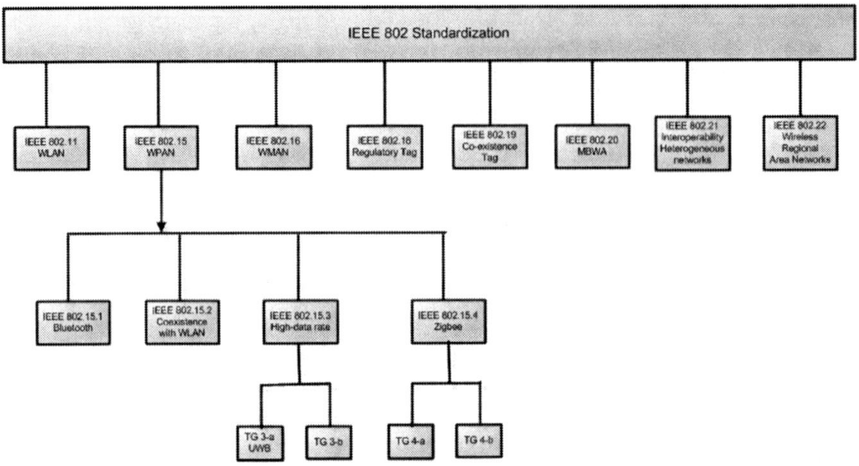

Fig. 10.6 The IEEE 802 standardization block diagram

layer for the WPAN high-speed short range applications. The IEEE 802.15.3 is chartered to draft and publish standard for low power low cost high-data rate (20 Mbps and higher) WPAN. In 2006 the 802.15.3 Task group TG-a has been withdrawn and the IEEE UWB standard creation has been pending upon commercially viable and proven technology [15]. The TG-b works on an amendment to 802.15.3 to improve the implementation and interoperability of the MAC. The IEEE 802.15.4 Zigbee was chartered to investigate a low data rate solution with multi-month to multi-year battery life and very low complexity. The Zigbee is an alliance of companies working together to enable low power wireless networks based on an open global standard. The IEEE 802.15.4 operates in an unlicensed, international frequency band. Potential applications are sensors, smart badges, remote controls and home automation. The task group TG4 of IEEE 802.15 has put itself into hibernation in March 2004 after forming the TG4.b [14]. The new task group has completed its work with the publication of revision on specific enhancements and clarifications of the 802.15.4 standard such as security key issues, consideration of newly available frequency allocations, and others. The IEEE 802.15.4-2006 standard was approved by IEEE standards board in 2006.

10.8 Summary

In this chapter we saw the efforts of various regulatory bodies across the globe to establish standards and norms towards UWB spectrum usage. The regulations, established as "mask", sets out upper limits on the amount of power that can be radiated at any particular frequency, both within and outside the core band of 3.1–10.6 GHz. The wireless networks of different countries operate at different frequencies and therefore the UWB regulation standards are formulated based on local needs. Therefore, devising a generic regulatory standard that caters to all markets across the globe will be one of the goals of the future.

Problems

Problem 10.1 While UWB is an unlicensed communication system why its standardization is important?

Problem 10.2 How do you think the location capability of UWB can influence the standardization of UWB?

Problem 10.3 Explain how the UWB market pressure could push the standardization of this technology?

Problem 10.4 What do you think about the battle of UWB high-speed short-range wireless standard against its alternative IEEE802.11n, especially in-home media networking market?

References

1. B. Allen (Editor), "Ultra Wideband:Technology and Future Perspectives," White paper, WWRF, v.3.0, March 2005.
2. http://ec.europa.eu/information_society/policy/radio_spectrum/by_topics/uwb/index_en.htm
3. http://europa.eu.int/information_society/policy/radio_spectrum/by_topics/uwb/index_en.htm
4. http://www.ero.dk/documentation/docs/doc98/official/Word/ECCDEC0604.DOC?frames = 0
5. B. Allen et al., "UWB: Applications, technology and future perspectives," International Workshop on Convergent Technologies, 2005.
6. online document at: www.onlinereporter.com/article.php?article_id = 8465
7. www.pulsers.eu
8. Online document at: www.swiss-ngn2005.org/posterPDFs/Swiss_Summit_WINDECT.pdf
9. W. Hirt et al., "Assessment of UWB radio regulatory matters: European and international developments," PULSERS Project, Deliverable D61.1b(i), Dec. 2005.
10. Online document http://techon.nikkeibp.co.jp/english/NEWS_EN/20060803/119881
11. www.aptsec.org
12. online document at: ieeexplore.ieee.org/iel5/9377/29767/01356311.pdf
13. ITU Web site: www.itu.org
14. http://standards.ieee.org/wireless/overview.html
15. http://ieee802.org/15/

Chapter 11
The Vision of Europe on UWB Applications

Contents

The European Commission (EC) is trying to push forward the boundaries of current radio technology looking at the next generation. The UWB is nowadays considered as an emerging technology and the work of EC is oriented to introduce the application of this in Europe with common rules, thus strengthening the internal market for information and communications technologies. The scope of the EU is to make UWB devices competitive in terms of costs compared to other markets, with the aim of avoiding the otherwise expected massive proliferation of equipment imported from other continents.

There are a lot of potential applications for the UWB technology and consequently a new emergently market. The main applications are Personal Area Networks (PAN) to link one person's devices together, or local area networks (LAN), to link devices in a room. This technology offers the opportunity of realizing very high data-rate for gigabit communication links, but it is also suitable for many other interesting applications, such as the low data rate links of the emerging Wireless Body Area Networks or pulse radars. For these reasons, the industrial and scientific communities are addressing many efforts for developing reliable, low-power consumption and low-noise UWB Radio Frequency Integrated Circuits (RFICs).

In the EU the research activities are organized into Framework Programmes (FP) for Research and Technological Development. The European UWB Cluster, which was established in the context of FP5 with the main objective of promoting interaction with European UWB regulation and standardisation bodies and exchange research results, promoted the development of UWB radio products in order to improve data transactions, localisation and tracking and low-cost systems and solutions but also in order to take advantage of its

H. Nikookar, R. Prasad, *Introduction to Ultra Wideband for Wireless Communications*, Signals and Communication Technology,
DOI 10.1007/978-1-4020-6633-7_11, © Springer Science+Business Media B.V. 2009

benefits for the economy. From then many different projects have been funding from the EU with this scope. Main projects dealing with UWB are:

UCAN (Ultra-wideband Concepts for Ad-hoc Networks) 01.01.2002–31.12.2004. It was a project with the objective to investigate the possibilities of an ultra-wideband (UWB) based physical layer (PHY) serving for an advanced, location based, self-organising medium access control (MAC) and network layer scheme and to provide a generic platform for a self-organising Wireless Personal Area Network.

ULTRAWAVES (Ultra Wideband Audio Video Entertainment System) 14.04.2001–13.10.2004. The objective was to provide a high performance and low cost wireless home connectivity solution, supporting applications requiring home multi streaming of high quality video and broadband multimedia.

There were other projects with indirect relation to UWB like as [1]:

- SAFETEL (Intelligent Integrated Safety System) to improve the robustness of motor vehicles against electromagnetic (EM) disturbances, using GPS, GSM, Bluetooth and Time Modulated UWB;
- U-R-SAFE a Personal Health Care system based on a Wireless Personal Area Network (WPAN) using UWB;
- PACWOMAN: Low-power, scalable and secure WPAN, using UWB;
- WIRENET a powerline data exchange for domestic and industrial automation based on UWB approach.

The main projects that deal with UWB and still are ongoing are Magnet and Pulsers, they will be described in next sections.

11.1 Magnet (My Personal Adaptive Global NET)

Target of the MAGNET project (01.01.2004–31.12.2005) was to develop a personal network (PN) that connects personal devices regardless of their location as in Fig. 11.1. Building block of the PN is a personal area network (PAN) that connects personal devices in close vicinity of the user. PNs consist of communicating clusters of personal digital devices shared with other people and may include infrastructure-based systems [2]. This is also one of the main applications of the UWB, and that is why mainly in the first phase of this project, a lot of studies have been done about the use of the UWB.

Objective of MAGNET was also to provide a highly adaptive and spectrally efficient PAN air-interface. It was envisioned to utilize, enhance and develop existing and upcoming air-interface technologies.

Previous projects investigated the feasibility of Multi Carrier (MC)-Techniques as a potential candidate for beyond 3G communications. MC-Techniques can accommodate advanced signal processing schemes that can enhance user throughput. Combinations of advanced signal processing techniques for MC-Techniques have been investigated, along with adaptive link layer techniques

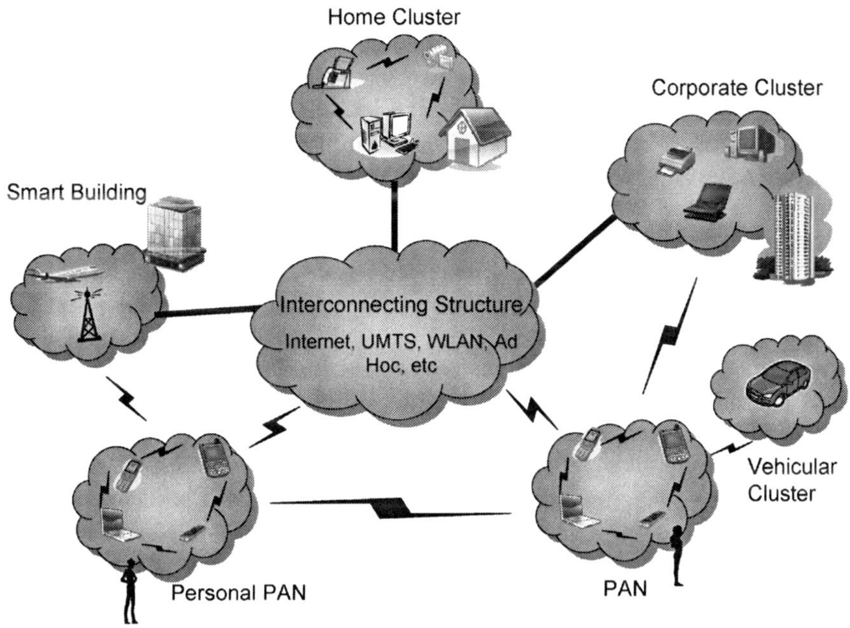

Fig. 11.1 Personal network concept

and layer interface issues [3]. Encouraging results have motivated interest into investigating the applicability of MC-Techniques for PAN networks, as a possible candidate for multiple-access for terminal-to-gateway connectivity. For the PAN gateway transceiver, additional schemes, like adaptive bit and power loading strategies and adaptive coding in combination with MIMO capabilities are mandatory for the flexibility and bandwidth efficiency required in future PAN, particularly for operation in the ISM Bands. Also the complexity of MC Air-Interfaces for PANs and their hardware feasibility especially with respect to processing power and power consumptions are critical research topics.

UWB techniques are well suited for short range communications, and have been addressed in several projects. The main peculiarity of MAGNET compared with others European projects was in the way that communication ranges from low data rates up to high data rate traffic (e.g. multimedia) for the same type of air-interface. Additionally, the PAN communication is limited to the immediate personal space (<10 m). As an alternative to the ISM Bands, the millimetre waveband shall offer similar bit rates as UWB in case of short-range systems. The very large available bandwidth (3 GHz on the up and down links) shall offer high bit rates up to 200 Mbit/s.

Before Magnet there was little evidence of measurements and modelling for UWB systems, and no studies focusing on PANs have been presented using any access method. In close proximity as well as close body environments, channel sounding has to take place in order to provide access to the mechanisms of the inherently (ultra)wideband channel and its dynamics. An appropriate and accurate channel model will take these into account and will be able to produce results that approximate real time varying multipath channels.

Interference of UWB systems or co-existence of any air-interface is a major issue and is currently being studied for fixed access mobile architectures. The ad hoc nature of the system proposed in MAGNET is innovative, and the study of the interference issues is required to ensure proper coordination as well as to minimise the impact on other systems operating in the same frequency bands.

However, the challenge to transfer and develop knowledge in the specific case of air-interfaces and inter-working of air-interfaces for PANs was to be tackled. The major issues was the investigation of the PAN channel and its characterisation, the investigation and development of access technologies tailored to the PAN including PHY Layer enhancement schemes and the development of adaptive MAC/RRM protocol suitable for the PAN environment as well as its inter-working with the PHY Layer and Upper Layers, requiring an overall cross-layer optimisation.

The MAGNET project at beginning identified three different kind of system [4], e.g. scenarios for low data rate (LDR) transmissions, medium data rate (MDR) transmission, high data rate (HDR) transmissions, as it possible to see in Table 11.1. In every system considered in Magnet the UWB technology has been evaluated as a possible solution. In the LDR system we expect to have a simple, low-cost, low-power consumption device allowing low data rate communications in a very short-range (approx. 0.5–5 m). In this context, the UWB technology seems to be a good candidate.

In the MDR it also expected low-complexity, low-power consumption devices allowing low and medium data rate communications in a short-range (approx. 1–5 m) scenarios.

For this scenario, UWB and MC technologies are good candidates to meet the requirements.

In the HDR system it is expected to have a more complex, greater power consumption device allowing medium to high data rate communications in a larger range than in the previous two scenarios (approx. 5–25 m).

Table 11.1 Scenarios in magnet [2]

Scenarlo	Class	Data rate (Mb/s)
LDR-WPAN	VLDR	<0.1
	LDR	<0.25
MDR-WPAN	MDR	<11
HDR-WPAN	HDR	<150
	VHDR	>150

The suggested technology for supporting these system requirements is a MC technique like for example OFDM, or in combination with CDMA, UWB, i.e. MC-CDMA or MC-UWB.

Within MAGNET different air-interface technologies as possible candidates have been identified and investigated. Five clusters have been formed to investigate the different options, namely MC-CDMA, OFDM-TDMA and MC-UWB for HDR transmissions, IR-UWB, OFDM-TDMA and MC-CDMA for MDR transmissions and FM-UWB and IR-UWB for LDR transmissions.

It is worth mentioning that the UWB technology has been considered in all LDR, MDR and HDR systems. The following sections provide a description of the UWB air-interfaces evaluated in the MAGNET.

11.1.1 MC-UWB

This air interface is particularly targeted to very high data rates up to 480 Mbps. In MAGNET two air interfaces have been investigated: one based on Multi Band OFDM (MBOA) operating in the [3.1–10.6] GHz frequency band and another one based as well on OFDM but operating at 60 GHz.

The results of the analysis show that the UWB performance evaluated in the [3.1–4.9] GHz exhibits a large sensitivity to the frequency selectivity due to the relative Cyclic Prefix (CP) duration with respect to the excess delay of the channel. This analysis has been carried out in applying a cut-off level S = [–10, –20 dB] on the impulse responses of the channel. When the excess delay of the channel is superior to the CP, ICI significantly damages performance while the frequency selectivity of the channel can be mitigated by an appropriate CP duration.

At 60 GHz, BER, has been compared for UWB and a wide band transmission modes. For each mode, it has been analysed the impact of the sub-carrier spacing on performance. Several spectrum efficiency schemes has been analysed. Furthermore, the impact of the binary interleaving depth on performance has been studied. UWB transmission mode takes benefit from the multipath channel in extreme cases (low selective and high selective multipath channels) proving that the millimetric UWB concept is a promising solution to VHDR WPANs. BER results and an analysis of the white noise power contribution to useful bits in [5] highlight the gains of high spectrum efficiency of UWB transmission at 60 GHz band.

11.1.2 FM-UWB

FM-UWB is a scalable air interface technology aimed at short-range (< 10 m) for LDR (up to 100 kbit/s) applications that is characterised by a low power consumption and ease-of implementation on an integrated circuit [5].

FM-UWB can be seen as an analogue implementation of a spread-spectrum system with a spreading gain equal to the modulation index β [6]. Frequency modulation has the unique property that the RF bandwidth is not only related to the FM bandwidth of the modulating signal, but also to the modulation index β that can be chosen freely. This yields either a bandwidth efficient narrow-band FM signal ($\beta < 1$) or a (ultra) wideband signal ($\beta >> 1$) that can occupy any required bandwidth compatible with the RF oscillator's tuning range. In the receiver processing gain is obtained by bandwidth reduction after the 1st FM demodulation.

A wideband delay line FM demodulator that is not preceded by any limiting amplifier constitutes the key component of the FM-UWB receiver. This unusual approach permits multiple users to share the same RF bandwidth up to the moment where the system capacity – like in a direct sequence spread spectrum system – is limited by the multiple-access interference.

The main advantages of FM-UWB may be summarised as [6]:

- ease of transmitter implementation;
- steep spectral roll-off of FM-UWB signal;
- robustness against narrowband jamming;
- no local oscillator required at the receiver;
- no carrier synchronisation needed;
- the system can be fully integrated.

The *transmitter* needs to comply with the FCC/ETSI spectral mask. A FM-UWB transmitter will not produce any energy below 3.1 GHz.

In [7] the FM-UWB LDR air interface and its performance in terms of BER with AWGN, multipath and interference have been investigated.

When operating in an AWGN channel, it was found that simulated and measured BER curves are close to the theoretical one, which means that the modelling of the complete FM-UWB system corresponds to reality. The AWGN BER curve was next used as reference to evaluate the system performance in a multipath channel and with both LDR IR and HDR Multi Band (MB)-OFDM interference.

It was found that channels with higher gain at the boundaries of the FM-UWB signal bandwidth give better performance. The effect of multipath upon a FM-UWB signal is mainly determined by the channel transfer function at the extremes of its bandwidth, corresponding to the extremes of the modulating sub-carrier signal that strongly influence the amplitude of the fundamental of the sub-carrier signal after the wideband FM demodulator. The values of the channel transfer function around the centre frequency hardly affect the performance since they correspond to the region around the zero crossing of the sub-carrier signal where its value is already close to zero.

The obtained BER curves show that the robustness of the FM-UWB in the presence of this interference in various channel propagation situations (BAN, PAN...). Furthermore, laboratory measurements performed with a LDR IR UWB interferer show that the BER degrades to 10^{-3} for a 10 dB stronger

IR-UWB interferer, which is coherent with the obtained simulation results. HDR MB-OFDM interference has also been simulated and good robustness figures were obtained; BER degrades to 10^{-3} for a 15 dB stronger MB-OFDM interferer.

11.1.3 IR-UWB

The UWB impulse radio (UWB-IR), is the traditional approach to the UWB for digital communication application involving the use of very short-duration pulses (0.1–1.5 ns) that occupy a single band of several GHz and usually the signal consists of a train of extremely narrow pulses with very low duty cycles.

A major advantage of IR-UWB is the conceptual simplicity of this approach to signal spreading.

In the context of the MAGNET project, IR UWB is proposed as a scalable air interface considered for the short range (to about 10 m) transmission case. The envisaged data rate transmission is Low Data Rate (LDR), potentially up to 10 Mbit/s. The main advantages of IR-UWB are the simplicity of implementation, the relaxed linearity requirements on amplifier design which enable the design of devices with ultra-low power consumption, potential resistance to interference due the inherent large effective spreading gain, and a high system capacity in terms of simultaneously supported users.

IR-UWB transmissions occupy a very broad frequency range, of which the upper and lower frequencies are unknown to parties not involved in the communication, coexistence that the coexistence with other system in operation is well addressed.

In [5] a detailed analysis of the optimum architecture of the IR UWB receiver was performed, its robustness to interference, its complexity and the coexistence aspects.

It was concluded that the case of a multipath channel, the windowed self multiplication received signal based procedure outperforms the pulse template generation based one. This is mainly due to two factors: the total cancellation of the interfering user (the user different from the user of interest) and the possibility of capturing a higher fraction of the signal energy which was dispersed due to the channel multipath, by using the selection window at the receiver. For highly dispersive channels (NLOS) the signal energy which can be captured in the observation window varies significantly, thus making the choice of optimal detector threshold more difficult.

It was also observed that in the multipath radio channels the inter-Time Hopping (TH) slot interference is the major factor influencing the BER performance of such a simple, non-coherent energy detector receiver, even in the single user scenarios. The result of a non appropriate TH code can be observed in a BER floor at high SNR values due to false detections, depending on the signal spread in the propagation channel. Hence, the TH codes have to be chosen appropriately in order to avoid inter-TH slot interference.

11.1.4 Conclusion on Interfaces

In the last phase of MAGNET project a decision has been taken on the interfaces that have to be implemented in MAGNET Beyond. The decision was to use only two interfaces, one for HDR and one for LDR. On the HDR side a PHY layer using Multi Carrier Spread Spectrum (MC-SS) techniques has been proposed for prototyping and this solution has been specified. The LDR system operates in ultra wide bands (UWB, 3.1 GHZ–10.6 GHZ) and utilises FM-UWB techniques along with an IEEE 802.15.4 MAC layer. For the FM-UWB PHY layer first projections on the power consumption were given.

11.2 Magnet Beyond

In MAGNET Beyond the interfaces selected in MAGNET have been implemented and a new requirement has been discussed: the simultaneous exploitation of the two air interfaces within one multimode device.

The necessity of the user terminal to transmit and receive towards and from different radio access networks requires a certain level of reconfigurability of the radio interface in order to allow the exploitation of different access technologies (e.g. MC-SS, FM-UWB, etc.) and different standards (e.g. IEEE 802.11a, IEEE 802.15.3 or IEEE 802.15.4) [8].

In this project the problem of interferences of the UWB with other interface has been coped with in order to integrate the two interfaces in the same device, as described in Table 11.2.

In the analysis of the interference between these two systems it has been shown that it can be classified as out-of-band interference. In fact considering Table 11.2 the major differences are the transmission power (34 dB of difference) and signal bandwidth. Signal attenuation as a function of distance is almost equal, since the two operating frequencies are relatively close to each other.

As the transmit power of HDR signals is 34 dB above the transmit power of LDR signals, and since the latter are additionally spread over a large frequency range, simultaneous LDR transmitters hardly have an impact on an HDR signal reception, except if interferer and receiver reside within one device.

Table 11.2 Characteristics of FM-UWB and MC-SS air interface [8]

Parameter	FM-UWB	MC-SS
Transmit Power	−14 dBm	+20 dBm
RF center frequency	4.5 GHz	5.25 GHz
RF signal bandwidth	500 MHz	36 MHz
RF signal envelope	Constant	Strongly varying
Predominant Modulation	FM	AM
Path loss @ 1 m	45 dB	46.5 dB

Table 11.3 Channel parameters for the investigated ultra wideband PAN/BAN scenarios [10]

User-proximity scenarios Parameters	PAN-FD	PAN-PD/ MD	BAN*
Log-normal wideband power shadowing std dev [dB]	3.0–4.5	2.1–3.6	1.5–3.0
Log-normal cluster fading std dev [dB]	4.0–5.6	2.2–6.9	4.3–5.4
Cluster power decay regions	<50 ns & >50 ns	<25 ns & >25 ns	<25 ns & >25 ns
Cluster power decay factor [dB/ns]	0.23 & 0.17	0.13 & 0.20	0.08 & 0.13
Average RMS delay spread [ns]	13–35	11–25	12–18
Average 90% energy delay window [ns]	25–88	20–57	24–49
Wideband power and signal clustering channel stationarity time	0.5	0.25	0.25

Note: *In these BAN scenarios 4 body-worn devices with *directional antenna* antenna and 1 belt-mounted handset with *omni directional antennas* have been used.

It has been found that LNA filtering, external filtering and antenna filtering are good solutions for the management of interference of HDR transmission on LDR reception. However, when the two air interfaces are located close to each other, MAC and higher layers mechanisms are needed to ensure coexistence.

Inspired by the recent evolution of UWB regulations in Europe, the operating frequency has evolved from the low band (3.1–5 GHz) to a limited portion of the low band (4.2–4.8 GHz) plus the high band (i.e. 6–9 GHz). This impacts the transmitter and receiver RF blocks which have become definitely more challenging to design.

The following scenarios have been investigated and are summarized in Table 11.3, with the UWB channel parameters:

- PAN-FD (Personal Area Network with Fixed Device)
- PAN-PD (Personal Area Network with Portable Device)
- PAN-MD (Personal Area Network with Mobile Device)
- BAN (Body Area Network)

More information about channel and RF specification can be found in [9].

11.3 Pulsers (Pervasive Ultra-Wideband Low Spectral Energy Radio Systems)

In Pulsers Phase I (2004–2005) [11] a new concept for short-range wireless systems based on the UWB-RT has been introduced. UWB-RT is capable of supporting wireless communication, ranging and localization applications. Considering data rate versus range, the technology accommodates two complementary classes of systems:

Systems offering high data rates (HDR) or very high data rates (VHDR) over links of up to a few metres. Systems supporting low data rates (LDR) alone or combined with location tracking (LDR-LT), covering distances up to tens of metres.

The main results of Pulsers phase I from the UWB point of view are the definition of user scenarios and business applications applying the unique technical properties of UWB-RT. Novel physical layer (PHY) concepts for HDR/VHDR and LDR-LT devices were developed and issues related to their interoperability explored and assessed.

PULSERS Phase II is an ongoing project (end in June 2008) with targets like:

(i) R & D on UWB-RT technology
(ii) Use scenario and business case evaluation, system concept development and integrated system definition inclusive the verification platform implementation
(iii) Contribution to regulation and standardisation enabling the use if UWB-RT.

For the LDR-LT UWB systems two different PHY have been proposed in [12]. These solutions are namely Pulse Position Modulation (PPM) energy detection and DBPSK 1-bit direct sub-sampling. The first solution (energy detection) is based on energy collection, which is mostly realised in analogue domain. The main characteristic of this architecture are its low complexity and straightforward implementation. The time position modulation is the preferred modulation and the transmitted signal is composed train of pulses.

The second solution proposed (1-bit direct sampling) combines differential demodulation and preamble detection schemes with a direct 1-bit subsampling of the incoming RF signal. This solution provides a good immunity against clock drift and also enables to maintain a reasonable clock speed. Furthermore the low power 1-bit quantization operation avoids the use of gain control mechanisms, provides fine immunity against interferes or near far effects, and tends to benefit from diversity on dense multipath profiles. The most significant part of the design complexity with UWB coherent systems is usually related to the synchronization functionality. Additional baseband modules were designed to perform the coherent integration of 1-bit signal prior to differential correlations, and the accumulation of soft differential correlation results for demodulation. There is also another solution based on energy detection that could possibly supplant the two main PHY solution described above. This solution can enable basic functions for different UWB pulse system specifications. The receiver is mainly based on energy detection with possibility to steer the detection window duration by the usage of an external reference clock [12].

In PULSERS II for Very High Data Rate UWB systems and evolution of the OFDM based WIMEDIA UWB standard has been developed. The main enhancement is the inclusion of higher order modulation modes up to 64-QAM for very short range communication needs and the increase of the basic operational bandwidth to 1 GHz.

These techniques will lead to two basic bandwidth modes: 528 MHz bandwidth with and without hopping, and 1056 MHz bandwidth without hopping [13].

11.4 Summary

The research community is developing UWB applications and investigating uses, standards and interference risks. FP5 and FP6 have been dealing with projects in this area and the UWB Cluster has the purpose to encourage UWB applications through the promotion, exchange and study of research projects' results. After the decision of the CEPT in March 2006, Europe have paved the way to ensure that Europe can utilize all the potential of UWB and compete in this market with USA and Asia.

References

1. European Commission, "Legal and Regulatory Constraints on the Application and Implementation of IST", January 2005.
2. MAGNET, "Requirement specification for PHY-Layer", IST-507102, My Personal Adaptive Global NET, deliverable D3.2.1, January 2004.
3. R. Prasad, J. Farserotu, K. Vandrup, "MAGNET paving a path towards the future wireless communication", Wireless Technology 2005.
4. MAGNET, "Requirement specifications for PHY layer.", IST-507102, My Personal Adaptive Global NET, deliverable D.3.2.1, July 2004.
5. MAGNET, "Candidate Air Interfaces and Enhancements", IST-507102, My Personal Adaptive Global NET, deliverable D3.2.2a, October 2004.
6. J. Gerrits and J. Farserotu, "Ultra wideband FM: A straightforward frequency domain approach", Conference European Microwave Week, Munich, Germany, October 2003.
7. MAGNET, "Update D3.2.2a, Candidate Air Interfaces and Enhancements", IST-507102, My Personal Adaptive Global NET, deliverable D3.2.2b, December 2005.
8. MAGNET Beyond, "Co-existence concept for Implementation of the FM-UWB and MC-SS RA Solutions", My Personal Adaptive Global NET and Beyond, D.3.3.1, June 2006.
9. MAGNET Beyond, "Prototype specification of antenna and radio front-end schemes for PAN devices", My Personal Adaptive Global NET and Beyond, D 3.1.1, June 2006.
10. MAGNET Beyond, "Prototype specification for the FM-UWB and MC-SS RA schemes", My Personal Adaptive Global NET and Beyond, D.3.2.1, June 2006.
11. http://www.pulsers.eu/
12. IST Pulsers, "LDR-LT concept specification – PHY and MAC layers", D 3a-3.1.
13. F. Berens, E. Dimitrov, T. Kaiser, A. Anttonen, A. Krause, A. Weir, "The PULSER II view towards very high data rate OFDM based UWB systems", Mobile and Wireless Communications Summit, 2007. 16th IST.

Index

Printed in the United Kingdom
by Lightning Source UK Ltd.
133868UK00005B/43/P

9 781402 066320